Jacqueline Stöckli

CD-MPR: Quo Vadis?

Jacqueline Stöckli

CD-MPR: Quo Vadis?

Sorting signals that regulate the cellular trafficking of the cation-dependent mannose 6-phosphate receptor (CD-MPR)

Südwestdeutscher Verlag für Hochschulschriften

Impressum/Imprint (nur für Deutschland/only for Germany)
Bibliografische Information der Deutschen Nationalbibliothek: Die Deutsche Nationalbibliothek verzeichnet diese Publikation in der Deutschen Nationalbibliografie; detaillierte bibliografische Daten sind im Internet über http://dnb.d-nb.de abrufbar.
Alle in diesem Buch genannten Marken und Produktnamen unterliegen warenzeichen-, marken- oder patentrechtlichem Schutz bzw. sind Warenzeichen oder eingetragene Warenzeichen der jeweiligen Inhaber. Die Wiedergabe von Marken, Produktnamen, Gebrauchsnamen, Handelsnamen, Warenbezeichnungen u.s.w. in diesem Werk berechtigt auch ohne besondere Kennzeichnung nicht zu der Annahme, dass solche Namen im Sinne der Warenzeichen- und Markenschutzgesetzgebung als frei zu betrachten wären und daher von jedermann benutzt werden dürften.

Coverbild: www.ingimage.com

Verlag: Südwestdeutscher Verlag für Hochschulschriften GmbH & Co. KG
Dudweiler Landstr. 99, 66123 Saarbrücken, Deutschland
Telefon +49 681 37 20 271-1, Telefax +49 681 37 20 271-0
Email: info@svh-verlag.de

Approved by: Basel, Universitaet Basel, Diss., 2003

Herstellung in Deutschland:
Schaltungsdienst Lange o.H.G., Berlin
Books on Demand GmbH, Norderstedt
Reha GmbH, Saarbrücken
Amazon Distribution GmbH, Leipzig
ISBN: 978-3-8381-2956-3

Imprint (only for USA, GB)
Bibliographic information published by the Deutsche Nationalbibliothek: The Deutsche Nationalbibliothek lists this publication in the Deutsche Nationalbibliografie; detailed bibliographic data are available in the Internet at http://dnb.d-nb.de.
Any brand names and product names mentioned in this book are subject to trademark, brand or patent protection and are trademarks or registered trademarks of their respective holders. The use of brand names, product names, common names, trade names, product descriptions etc. even without a particular marking in this works is in no way to be construed to mean that such names may be regarded as unrestricted in respect of trademark and brand protection legislation and could thus be used by anyone.

Cover image: www.ingimage.com

Publisher: Südwestdeutscher Verlag für Hochschulschriften GmbH & Co. KG
Dudweiler Landstr. 99, 66123 Saarbrücken, Germany
Phone +49 681 37 20 271-1, Fax +49 681 37 20 271-0
Email: info@svh-verlag.de

Printed in the U.S.A.
Printed in the U.K. by (see last page)
ISBN: 978-3-8381-2956-3

Copyright © 2011 by the author and Südwestdeutscher Verlag für Hochschulschriften GmbH & Co. KG and licensors
All rights reserved. Saarbrücken 2011

Table of Contents

SUMMARY 3

GENERAL INTRODUCTION 5

PART I:
THE PALMITOYLTRANSFERASE OF THE CATION-DEPENDENT MANNOSE 6-
PHOSPHATE RECEPTOR CYCLES BETWEEN THE PLASMA MEMBRANE AND
ENDOSOMES 67

PART II:
THE ACIDIC CLUSTER OF THE CD-MPR BUT NOT PHOSPHORYLATION IS
REQUIRED FOR GGA1 AND AP1 BINDING 90

GENERAL DISCUSSION 110

REFERENCES 119

Summary

Lysosomes are membrane-bound organelles that serve in the degradation of many extracellular and intracellular macromolecules. Lysosomal biogenesis depends on the delivery of newly synthesized lysosomal hydrolases. This process requires the acquisition of the lysosomal targeting signal, the mannose 6-phosphate tag that is specifically recognized by mannose 6-phosphate receptors (MPRs) in the TGN. The receptor-ligand complex is subsequently packaged into clathrin-coated vesicles and transported to early endosomes. The lower pH in the endosomal compartment causes the dissociation of the MPR and the ligand. The lysosomal enzymes are transferred to the lysosome, where they are activated, whereas the MPRs are transported from endosomes back to the TGN where they mediate another round of transport. Two distinct MPRs were identified and characterized - the 46 kDa cation-dependent (CD) MPR and the ~300 kDa cation-independent (CI) MPR. This study concentrates on the CD-MPR.

The intracellular trafficking of the CD-MPR is mediated by sorting signals located in its cytoplasmic tail of 67 amino acids. The sorting motifs are recognized by specific adaptor proteins that mediate the vesicular transport of the receptor. Although several motifs and their interacting partners were identified in the CD-MPR, the various trafficking steps are not yet fully understood. In this study we focused on the characterization of two motifs of the receptor - the cysteine C^{30} and C^{34} which undergo reversible palmitoylation and the acidic cluster of the casein kinase 2 (CK2) phosphorylation site (E^{55}-E^{56}-S^{57}-E^{58}-E^{59}).

The CD-MPR is transported efficiently from late endosomes back to the TGN since only a very small percent of receptors are missorted to lysosomes where they are rapidly degraded. This transport step depends on the palmitoylation of C^{34}, and additionally on the diaromatic motif $F^{18}W^{19}$. The membrane anchoring mediated by the palmitate, 34 amino acids away from the transmembrane domain, implies a drastic conformational change on the cytoplasmic tail of the CD-MPR. The diaromatic motif is likely to be better exposed to the interacting protein in the palmitoylated than in the non-palmitoylated CD-MPR. Our hypothesis suggests that the reversible palmitoylation regulates the sorting signals in the cytoplasmic tail of the receptor. This would require that the palmitoylation occur enzymatically. In Part I, we show that indeed the palmitoylation depends on a membrane-bound enzyme. This palmitoyltransferase cycles between the plasma membrane and endosomes. Close proximity of the palmitoyltransferase to the site where the palmitoylation of the CD-MPR is required is optimal to ensure the presence of the palmitoylated C^{34} in late endosomes. Thus, the localization of the palmitoyltransferase supports our hypothesis of palmitoylation as a regulatory mechanism for the sorting signals in the cytoplasmic tail of the receptor.

Summary

Correct sorting of the CD-MPR from the TGN to endosomes depends on the D^{61}-X-X-L^{64}-L^{65} sequence, which interacts with GGA (Golgi-localizing, γ-ear-containing, ARF-binding protein), a monomeric adaptor protein that mediates the formation of clathrin-coated vesicles at the TGN. Several substrates of GGA have a CK2 site upstream of the DXXLL motif and in two cases, phosphorylation by CK2 was shown to increase the affinity of GGA1 to cargo. The CD-MPR also contains a CK2 site upstream of the DXXLL motif, but its involvement in GGA1 binding has not been investigated so far. The CK2 site of the CD-MPR was shown to interact with the adaptor protein 1 (AP-1), another protein involved in the sorting of cargo in the TGN, possibly in cooperation with GGA. Previous reports on the requirement of phosphorylation of the CD-MPR for binding to AP-1 were controversial. In Part II, we analyzed the influence of the CK2 phosphorylation site of the CD-MPR in binding to GGA1 and AP-1 and thus, in sorting in the TGN. A mutational analysis revealed that high affinity binding between CD-MPR and GGA1 was dependent on the acidic amino acid E^{59} and to a lesser extent on E^{58}, while the phosphorylation of the S^{57} had no influence, indicating that the GGA1 binding site in the CD-MPR extends to E^{58}-E^{59}-X-D^{61}-X-X-L^{64}-L^{65}. In contrast, AP-1 depended on all glutamates surrounding the serine E^{55}, E^{56}, E^{58}, E^{59} in the CD-MPR for binding, but was also independent of the phosphorylation of S^{57}. Therefore, we revealed that the phosphorylation of S^{57} is not required for sorting in the TGN. Interestingly, the binding affinity of GGA1 to the CD-MPR was 2.4-fold higher than that of AP-1 to the partially overlapping binding site in the CD-MPR. Thus, we present a modified model for the sorting process in the TGN, involving both GGA1 and AP-1, where the different binding affinities, determine the order of binding to the partially overlapping binding sites in the CD-MPR. First, GGA1 binds to the CD-MPR due to its higher affinity and is subsequently released from the CD-MPR as a result of its autoinhibition caused by phosphorylation. This allows the AP-1 to bind and recruit the remaining components for correct sorting of the CD-MPR in the TGN.

With our work we contributed to the understanding of specific transports steps of the CD-MPR and thereby we are advancing towards the goal of fully elucidating the trafficking of the receptor.

General Introduction

ABBREVIATIONS 7

1 PROTEIN TRAFFICKING 9

    1.1 Proteins involved in trafficking 9
        1.1.1   Clathrin 9
        1.1.2   Adaptor proteins 11
        1.1.3   SNAREs and Rabs 12

    1.2 Clathrin-mediated endocytosis 13
        1.2.1   Proteins involved in clathrin-mediated endocytosis 14
        1.2.2   Acting together for vesicle formation 19

    1.3 Sorting in the TGN 20
        1.3.1   Transport from the TGN to endosomes 21
        1.3.2   Sorting into secretory granules 27

    1.4 Sorting in endosomes 29
        1.4.1   Formation of multi-vesicular bodies / late endosomes 29
        1.4.2   Transport from endosomes to the TGN 31

2 MANNOSE 6-PHOSPHATE RECEPTOR 36

    2.1 Structure and biosynthesis of MPRs 36
        2.1.1   The CD-MPR 36
        2.1.2   The CI-MPR 42
        2.1.3   Comparison between the CD-MPR and the CI-MPR 46

    2.2 Function and relevance of MPRs 46
        2.2.1   Lysosomal biogenesis 46
        2.2.2   Role of CI-MPR at the cell surface 49
        2.2.3   Ligands of the MPRs 50
        2.2.4   MPR-independent lysosomal targeting 50
        2.2.5   Lysosomal storage disorders 50

3 PALMITOYLATION 53

    3.1 Lipid modifications 53

3.1.1	N-Myristoylation	53
3.1.2	Prenylation	55
3.1.3	Palmitoylation	56
3.1.4	Dual fatty acylation	56

3.2 Palmitoylation motif of proteins 57

3.3 Palmitoyltransferases 59

3.4 Intracellular sites of palmitoylation 60

3.5 Palmitoylthioesterases 60

3.6 Role of palmitoylation 61

3.6.1	Palmitoylation in signal transduction	61
3.6.2	Palmitoyation for localization to lipid rafts	63
3.6.3	Palmitoylation in protein trafficking	64

4 AIM OF THE THESIS 65

Abbreviations

AP-1, -2, -3, -4	adaptor protein 1, 2, 3, 4
APT1	acylprotein thioesterase 1
ARF1	ADP-ribosylation factor 1
BFA	brefeldin A
CALM	clathrin assembly lymphoid myeloid leukaemia protein
CCV	clathrin-coated vesicle
CD-MPR	cation-dependent mannose 6-phosphate receptor
CHC	clathrin heavy chain
CI-MPR	cation-independent mannose 6-phosphate receptor
CK2	casein kinase 2
CLC	clathrin light chain
CRD	cysteine-rich domain
Dab2	disabled 2
EE	early endosome
EEA1	early-endosomal autoantigen 1
EGFR	epidermal growth factor receptor
EH	Eps15 homology
ENTH	epsin N-terminal homology
Eps15	EGFR-pathway substrate 15
Eps15R	Eps15 related
epsin1	Eps15 interacting protein 1
epsinR	epsin-related
ER	endoplasmic reticulum
ERGIC	ER-Golgi intermediate compartment
ESCRT-I, -II, -III	endosomal complexes required for transport-I, -II, -III
FT	farnesyltransferase
FYVE domain	conserved in Fab1p/YOTB/Vac1p/EEA1
GAE	γ-adaptin ear
GAK	G-cyclin-associated kinase
GAP	GTPase activating protein
GAT	GGA and TOM1
GDF	GDI-displacement factor
GDI	GDP-dissociation inhibitor protein
GED	GTPase effector domain
GEF	guanine nucleotide exchange factor
GFP	green fluorescence protein
GGA1, 2, 3	Golgi-localizing, γ-ear-containing, ARF-binding protein 1, 2, 3
GGT-1, -2	geranylgeranyltransferase 1, 2
GlcNAc	N-acetylglucosamine
GPCR	G-protein coupled receptor
G-protein	guanine-nucleotide-binding protein
HA	influenza virus hemagglutinin A
HIP1	huntingtin interacting protein 1
HIP1R	HIP1 related
Hrs	hepatocyte growth factor-regulated tyrosine kinase substrate
ICD	I-cell disease
IGF-II	insulin-like growth factor II
ISG	immature secretory granules
ITAM	immunoreceptor tyrosine-based activation motif

LAMP-1, -2	lysosomal-associated membrane protein 1, 2
LAT	linker for activation of T cells
LBPA	lyso-bisphosphatidic acid
LE	late endosome (also called MVB)
LIF	leukemia inhibitory factor
LIMPII	lysosomal integral membrane protein II
LPA	lyso-bisphosphatidic acid
M6P	mannose 6-phosphate
MARCKS	myristoylated alanine-rich C kinase substrate
MPR	mannose 6-phosphate receptor
MSG	mature secretory granules
MVB	multi-vesicular body
NMT	N-myristoyl transferase
NSF	N-ethylmaleimide-sensitive factor
PACS-1	phosphofurin acidic cluster sorting protein 1
PC6B	proprotein convertase 6B
PH	pleckstrin homology
phosphotransferase	UDP-GlcNAc:lysosomal enzyme GlcNAc-1-phosphotransferase
PI(3)K	PtdIns(3)-kinase
PKC	protein kinase C
PNS	post-nuclear supernatant
PPT1, 2	palmitoylthioesterase 1, 2
PRD	proline-rich domain
PTB	phosphotyrosine-binding
PtdIns(3)P	phosphatidylinositol-3-phosphate
PtdIns(4,5)P_2	phosphatidylinositol-4,5-bisphosphate
PX	phox homology
Rab	Ras-like in rat brain
RGS	regulator of G-protein signaling
Sac1	suppressor of actin 1 (phosphatase activity)
SH3	Src-homology 3
SNARE	soluble NSF attachment protein receptor
SNX	sorting nexin
STAM1, 2	signal-transducing adaptor molecule 1, 2
TGF-β	transforming growth factor-β
TGN	trans-Golgi network
TIP47	MPR tail interacting protein of 47 kDa
TMD	trans-membrane domain
Tsg101	tumor-susceptibility gene product 101
UCE	"uncovering" enzyme (GlcNAc-1-phosphodiester α-N-acetylglucosaminidase)
UIM	ubiquitin-interacting motif
VHS	conserved in Vps27, Hrs, STAM
VSV-G	vesicular stomatitis virus G
α-SNAP	α-soluble NSF attachment protein
$β_2$AR	$β_2$-adrenergic receptor

1 Protein trafficking

The biogenesis of the different organelles in the cell depends on the proper delivery of the proteins exhibiting an essential function required by the specific organelle. The proteins are synthesized and inserted into the ER membrane and are transported through many organelles via vesicular transport steps to the final destination. In addition to the biosynthetic pathway, the endocytic pathway takes up proteins from the extracellular space and the plasma membrane through endocytosis (see Figure 1). Ligand-binding receptors at the plasma membrane internalize and deliver the ligands for lysosomal degradation, whereas the receptors either recycle or are down-regulated by lysosomal degradation. The proper operation of these transport routes requires several important decisions to be made along the way. At the plasma membrane, proteins either remain at the cell surface or are rapidly internalized into endosomes. At the TGN, the choice is between going to the plasma membrane and being transported to endosomes. In endosomes, proteins can either recycle to the plasma membrane or the TGN or go to lysosomes. These decisions are mediated by a complex system of sorting signals in the proteins and a molecular machinery that recognizes those signals and delivers the proteins to their intended destinations.

1.1 Proteins involved in trafficking

For a specific transport step many components are required. A certain sorting signal in a protein destined for transport to a specific organelle is only recognized in the correct environment produced by the lipid composition of the membrane of the organelle and by accessory proteins recruited to the membrane. A soluble cytoplasmic protein, upon recognizing the signal in cargo, recruits other proteins to the budding site. These proteins might in turn recruit even more proteins, resulting in a complex system that is required for a vesicle to bud on a certain donor organelle. A general mechanism for all transport steps was elucidated, involving "adaptor proteins", which recognize the signal and recruit "coat proteins" which coat the vesicle (Schekman and Orci, 1996). The process requires many "accessory proteins" with a function in facilitating vesicle formation, coat assembly or separating the vesicle from the membrane. Further specificity is endowed by the SNARE and Rab proteins, which direct the vesicles to the correct donor organelle and facilitate the fusion. One component of the vesicles that is not specific for one organelle and is present on numerous vesicles budding from different organelles is the coat protein clathrin.

1.1.1 Clathrin

Clathrin is composed of three 192 kDa heavy chains (CHC), each bound to either of the two ~30 kDa light chains (CLCa or CLCb). This complex is called a triskelion for its three-legged appearance. Triskelions are the assembly units of the polygonal lattice composed of hexagons and pentagons formed at the bud site and eventually pinched off from the membrane thus enclosing a vesicle (Pearse, 1976). It is suggested that the clathrin coat initially forms a flat network of hexagons, some of which are then converted into pentagons, thus driving the curvature of the membrane (Heuser, 1980). Whether it is the clathrin or the membrane curvature that induces the conversion of some hexagons into pentagons is not known.

Clathrin itself does not interact with the membrane directly. It is recruited to the membranes through binding to an adaptor protein (Vigers et al., 1986). Several clathrin binding motifs were identified in adaptor proteins, the $L\Phi X\Phi[D,E]$ motif (where Φ is a bulky hydrophobic amino acid), also referred to as the clathrin box, the PWDLW sequence, the LLDLL sequence and the short DLL repeats. All of these motifs bind to the N-terminal domain of the CHC (Dell'Angelica et al., 1998;

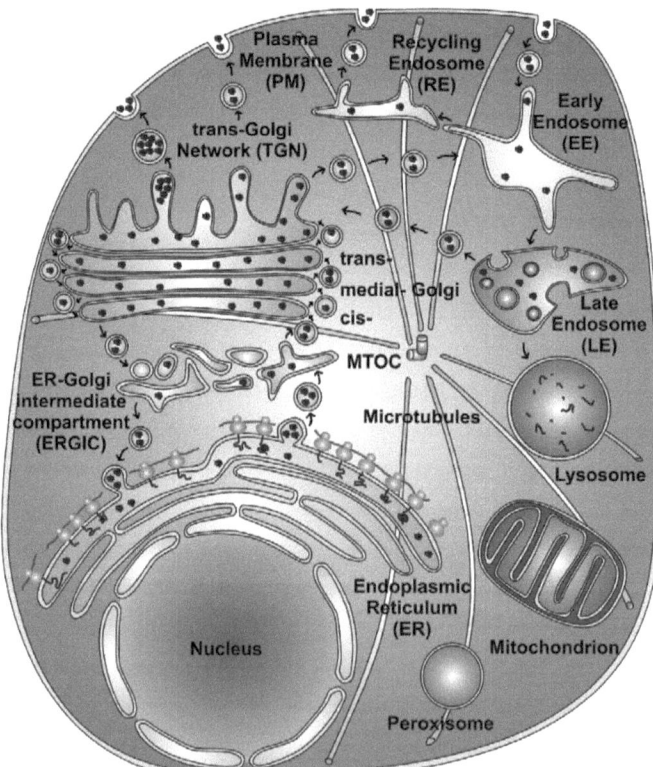

Figure 1: Intracellular Trafficking. A eukaryotic cell is illustrated including the various organelles. The organelles are not drawn to scale. The numerous possibilities of the protein trafficking are depicted for a newly synthesized protein, targeted into the lumen of the ER. The individual transport steps are marked with black arrows. The ribosomes localized on the rough ER membrane are in green, the lumenal proteins are blue. MTOC, microtubule organizing center.

Ramjaun and McPherson, 1998; Kirchhausen, 2000; Morgan et al., 2000; Doray and Kornfeld, 2001). The specificity of clathrin-coated vesicles originating from the TGN, the plasma membrane and endosomes, is mediated by the adaptor proteins. The adaptor proteins in turn are recruited to the membranes by interaction either with a small GTPase or with lipids, which are specific for a certain organelle. There are several kinds of clathrin-coat adaptor proteins, adaptor protein-1 (AP-1), AP-2, GGA (Golgi-localizing, γ-ear-containing, ARF-binding protein) and Hrs (hepatocyte growth factor-regulated tyrosine kinase substrate) which mediate formation of clathrin-coated vesicles at different organelles (Bonifacino and Lippincott-Schwartz, 2003). AP-1 and GGA both are recruited to the TGN through interaction with ARF1 and might be part of the same clathrin coat involved in sorting from the TGN to endosomes (Stamnes and Rothman, 1993; Puertollano et al., 2001b). AP-2 is recruited to the plasma membrane by binding to phosphatidylinositol 4,5-bisphosphate (PtdIns(4,5)P$_2$) and mediates clathrin-dependent endocytosis (Collins et al., 2002). Hrs is recruited to early endosomes by binding to phosphatidylinositol 3-phosphate (PtdIns(3)P) through its FYVE domain (conserved in Fab1p/YOTB/Vac1p/EEA1) and is involved in intralumenal invagination to form intralumenal vesicles in multi-vesicular bodies (Raiborg et al., 2001b; Raiborg et al., 2002).

There are also non-clathrin coats involved in certain intracellular trafficking steps such as the coatamer protein I (COPI), COPII, AP-3 and AP-4. COPI mediates retrograde transport from the Golgi and pre-Golgi compartments to the ER (Ostermann et al., 1993). COPII is involved in anterograde transport from the ER to Golgi (Barlowe et al., 1994). AP-3 targets lysosomal membrane proteins to lysosomes and lysosome-related organelles (Le Borgne et al., 1998). Although the AP-3 binds to clathrin *in vitro*, it is not required for the function of AP-3, which is therefore suggested to operate without clathrin (Simpson et al., 1997). AP-4 is probably involved in sorting from the TGN to the basolateral plasma membrane (Dell'Angelica et al., 1999a).

1.1.2 Adaptor proteins

The adaptor proteins provide the link between the coat and the cargo and are responsible for recruiting accessory proteins. There are two groups of adaptor proteins, the heterotetrameric complexes, named adaptor proteins 1 to 4 (AP-1, AP-2, AP-3 and AP-4) and the monomeric adaptors, GGAs and Hrs.

The adaptor protein (AP) complexes are a family of heterotetrameric complexes consisting of four adaptins. Each complex is composed of two large adaptins (one each of γ/α/δ/ε and β1-4, respectively, of 90-130 kDa), one medium adaptin (μ1-4, of ~50 kDa) and one small adaptin (σ1-4, of ~20 kDa) (Robinson and Bonifacino, 2001). The hinge domains of the β adaptins of AP-1 and AP-2 interact with clathrin through their clathrin boxes. AP-2 contains an additional clathrin binding site in the β2 ear domain (Kirchhausen, 2000; Owen et al., 2000). The μ and β adaptins are implicated in cargo selection, whereby the μ subunit binds to YXXΦ motifs and the μ or β subunit binds to dileucine motifs (Ohno et al., 1995; Owen and Evans, 1998; Rapoport et al., 1998). This binding to YXXΦ motifs is a feature of the μ subunits of all four AP complexes with each μ adaptin recognizing a distinct but overlapping set of YXXΦ signals (Ohno et al., 1998; Aguilar et al., 2001). The ears of the α, β and γ subunits recruit accessory proteins that participate in events such as vesicle scission and vesicle uncoating (Slepnev and DeCamilli, 2000). The individual APs and their accessory proteins are described in more detail in the chapters: *Clathrin-mediated endocytosis* and *Transport from the TGN to endosomes*.

The other adaptor proteins involved in vesicle formation at the TGN (GGA1, GGA2, GGA3) and in endosomes (Hrs) are monomeric proteins. They are composed of several domains and motifs, including a domain that targets them to the specific membranes, a clathrin box, and a domain that interacts with cargo. A detailed description of these adaptor proteins and their accessory proteins is found in the chapters: *Transport from the TGN to endosomes* and *Formation of multi-vesicular bodies*.

1.1.3 SNAREs and Rabs

Additional proteins are required to generate specificity in targeting and fusion of the vesicles with the appropriate acceptor membranes. Two classes of proteins have emerged as essential players in many vesicle transport processes. The SNARE (soluble N-ethylmaleimide sensitive factor adaptor protein receptor) family is necessary for docking and fusion and the Rab (Ras-like in rat brain) family is required for several steps in the vesicle transport, including the tethering of the vesicle prior to the function of the SNAREs.

SNAREs are membrane proteins that are localized to various intracellular organelles. SNAREs contain a characteristic heptad repeat sequence known as the SNARE motif. The synaptic SNARE complex, composed of VAMP, syntaxin and SNAP-25, was the first complex identified and most current ideas of SNARE function are based on this model (Sollner et al., 1993). Based on the localization of the SNARE they are termed v-SNARE for vesicle localized and t-SNARE for target membrane localized. The specificity of vesicle docking and fusion is determined by proper v-SNARE:t-SNARE interactions. A newer classification system has been proposed based on whether a conserved glutamine (Q) or arginine (R) is present in the SNARE motif, although Q-SNAREs are almost always t-SNAREs and R-SNAREs are almost always v-SNAREs (Fasshauer et al., 1998). Characterized SNARE complexes always contain three Q-helices (collectively, the t-SNARE) and one R-helix (the v-SNARE), although these may be contributed from three or four proteins (SNAP-25 homologs typically contribute 2 helices). Syntaxins (t-SNAREs) interact with Sec1-like proteins that act as chaperones preventing SNARE complex assembly until signaled to release syntaxin

(Jahn, 2000). After dissociation from the Sec1-like protein, syntaxin interacts with VAMP (v-SNARE) and SNAP-25 (t-SNARE) forming a coiled-coil bundle (Poirier et al., 1998). This formation brings the vesicle and target membrane together, resulting in fusion (Weber et al., 1998). Two soluble proteins, NSF (N-ethylmaleimide sensitive factor) and α-SNAP (α-soluble NSF attachment protein) act to disassemble SNARE complexes. α-SNAP targets the ATPase NSF to the SNARE complex by interacting with the SNAREs and stimulates the ATPase which is essential for the SNARE complex disassembly (Morgan et al., 1994).

Rab proteins comprise the largest family within the Ras superfamily of small GTPases. Rabs are involved in the regulation of intracellular transport steps and are implicated in the control of vesicle docking and fusion (Gonzalez and Scheller, 1999). In humans, more than 60 distinct Rab proteins have been identified and each is believed to be associated with a particular organelle or pathway (Bock et al., 2001). The Rabs are prenylated by geranylgeranyltransferase (GGT2), which helps them to anchor to the membrane (Desnoyers et al., 1996). However, only the GTP-bound form of Rabs is membrane bound. Rabs cycle between an inactive GDP-bound and an active GTP-bound conformation (Rybin et al., 1996). The GDP/GTP exchange reaction is catalyzed by guanine nucleotide exchange factors (GEFs), such as Rabex-5 for Rab5 (Horiuchi et al., 1997). In the GDP-bound form, Rabs bind to GDP-dissociation inhibitor (GDI) proteins, which are released by a GDI-displacement factor (GDF) upon GDP/GTP exchange (Dirac-Svejstrup et al., 1997). A GTPase activating protein (GAP) then stimulates the GTP hydrolysis.

Rabs have a role in multiple aspects of vesicular transport. Involvement of Rabs in the formation of vesicles was suggested by *in vitro* studies in mammalian cell lines (Gorvel et al., 1991; McLauchlan et al., 1998). Secondly, a role for Rabs in vesicle motility has been suggested by the finding that Rab5 stimulates both endosome association with, and movement along microtubules (Nielsen et al., 1999). Thirdly, through the interaction of Rab5 with PI(3)-kinase (PI(3)K), which generates PtdIns(3)P, a role for Rabs in membrane remodeling has been proposed (Christoforidis et al., 1999). The role of Rabs in vesicle docking is suggested to involve tethering of adjacent membranes prior to their fusion (Waters and Pfeffer, 1999). Finally, Rabs might play a role in membrane fusion itself via regulation of SNARE complex formation. Rabenosyn-5, a Rab5 effector, interacts with a Sec1-like protein involved in SNARE complex formation (Nielsen et al., 2000).

1.2 *Clathrin-mediated endocytosis*

Clathrin-mediated endocytosis is involved in the internalization of receptors and extracellular ligands, for the recycling of plasma membrane components and for the retrieval of surface proteins destined for degradation. At the plasma membrane, selected cargo is recruited to a coat, assembled through the involvement of adaptor protein AP-2 and polymerization of clathrin, a process assisted by accessory proteins. The proteins involved form a complex endocytosis machinery (see Figure 2).

1.2.1 Proteins involved in clathrin-mediated endocytosis

1.2.1.1 Adaptor and "adaptor-like" proteins

Adaptor protein-2 (AP-2)

AP-2 is composed of the subunits α1 or α2, β2, μ2 and σ2 (Robinson and Bonifacino, 2001). AP-2 is recruited to the plasma membrane by interacting with PtdIns(4,5)P$_2$, a phospholipid enriched at the plasma membrane, through binding sites in the μ2-subunit and in the trunk region of

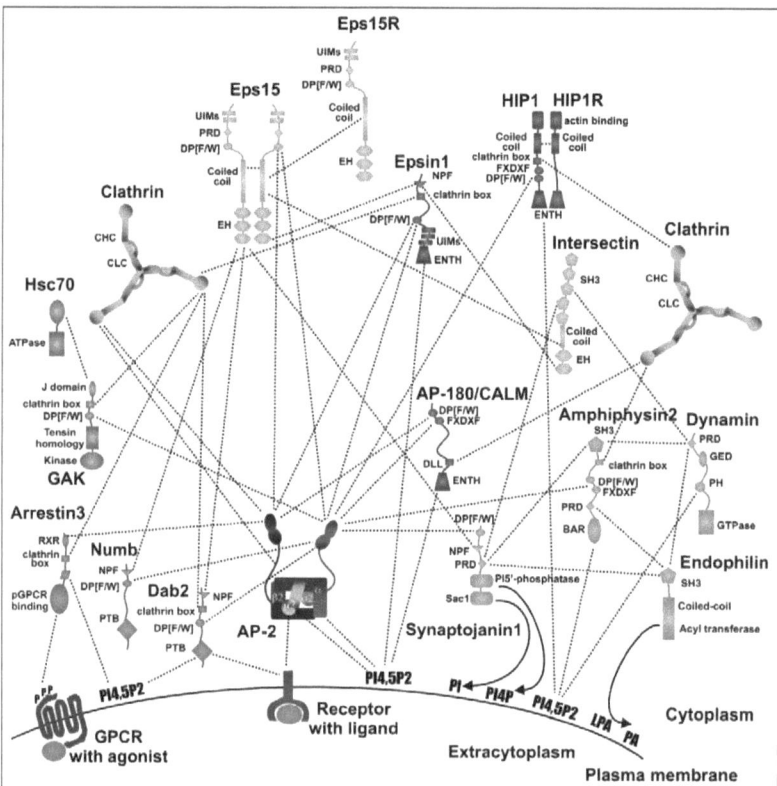

Figure 2: AP-2 and clathrin with accessory proteins. The domain structure of the proteins is not drawn to scale. The proteins are shown in different colours according to their function in vesicle formation and budding. The alternative adaptor proteins are in red, the ENTH-containing proteins are in purple, EH-containing proteins are in green, proteins involved in the fission process from the membrane are in orange and proteins involved in the dissociation of clathrin after budding are in brown. The interactions are depicted by dotted lines. The names of the domains and the proteins are found in the text.

α-subunit (Stauffer et al., 1998; Collins et al., 2002). AP-2 interacts with clathrin not only through the β2 hinge domain, but also the β2 ear domain (Kirchhausen, 2000; Owen et al., 2000). The μ2 subunit of AP-2 interacts with the FXNPXY motif or the YXXΦ motif in cargo, such as LDL receptor, transferrin receptor, epidermal growth factor receptor (EGFR), and TGN38 (Ohno et al., 1995; Boll et al., 1996; Boll et al., 2002). In addition, [D,E]XXXL[L,I] motifs (e.g. in the invariant chain) are recognized by AP-2 either through the μ2 or the β2 subunit (Bremnes et al., 1998; Rapoport et al., 1998). However, overexpression of a protein containing an YXXΦ motif leads to mislocalization of other proteins containing an YXXΦ motif, but not of proteins containing a [D,E]XXXL[L,I] motif and vice versa, indicating separate but saturable binding pockets in AP-2 (Marks et al., 1996). Phosphorylation of the tyrosine 156 of μ2 by the adaptor-associated kinase (AAK1) is required for efficient cargo binding (Ricotta et al., 2002). AP-2 sorts the cargo into clathrin-coated pits and binds to many accessory proteins that either facilitate vesicle formation or are involved in the release of the vesicle from the membrane. Most accessory proteins interact with the α adaptin ear of AP-2, such as Eps15 (EGFR-pathway substrate 15), epsin1 (Eps15 interacting protein 1), Dab2 (disabled-2), numb, AP-180/CALM (clathrin assembly lymphoid myeloid leukaemia protein), synaptojanin1, amphiphysin2, HIP1 (huntingtin interacting protein 1) and GAK (G-cyclin-associated kinase) (Benmerah et al., 1996; David et al., 1996; Owen et al., 1999; Slepnev and DeCamilli, 2000; Brett et al., 2002). The motifs in the accessory proteins that interact with the α adaptin ear were identified to be the DP[F/W] motif or the FXDXF motif. Some DP[F/W] motif containing accessory factors such as Eps15, epsin1 and AP-180/CALM, were also shown to interact with the β2 adaptin ear (Owen et al., 2000). However, this binding site overlaps with the clathrin binding site in the β2 adaptin ear. Thus, the polymerization of clathrin, which is promoted by interacting with β2 hinge and ear, would cause the controlled release of accessory proteins at sites of vesicle formation.

"Adaptor-like" proteins

Some of the accessory proteins that interact with AP-2 and clathrin bind to cargo as well. These proteins include arrestin3 (also called β-arrestin2) and two phosphotyrosine-binding domain (PTB) containing proteins, disabled-2 and numb.

Arrestin3 is involved in G-protein-coupled receptor (GPCR) internalization and degradation (Miller and Lefkowitz, 2001). Arrestin3 interacts with the $β_2$-adrenergic receptor ($β_2$AR) upon phosphorylation of the receptor cytoplasmic domain caused by agonist stimulation of the $β_2$AR. Arrestin3 also interacts with AP-2, clathrin and the E3 ubiquitin ligase Mdm2 (Shenoy et al., 2001). Upon binding to the $β_2$AR, both the arrestin3 and the $β_2$AR are mono-ubiquitinated by Mdm2. Mono-ubiquitination of arrestin3 is essential for internalization of the receptor-arrestin3 complex. In contrast, mono-ubiquitination of the receptor is not essential for internalization but for subsequent sorting in endosomes for degradation (see also chapter: *Formation of multi-vesicular bodies*) (Shenoy et al., 2001). Following ubiquitination, the β_2AR is internalized in complex with arrestin3, AP-2 and clathrin. Since two accessory proteins which interact with AP-2, contain an ubiquitin-interacting motif (UIM), they might be possible binding partners of the ubiquitinated arrestin3, thereby facilitating endocytosis. Arrestin3 represents a monomeric adaptor, which might target GPCR for internalization, by interacting with the AP-2/clathrin endocytosis machinery.

The phosphotyrosine binding domain (PTB) in disabled-2 (Dab2) and numb bind to FXNPXY sequences, however, the name of the domain is misleading since both have a preference for non-phosphorylated tyrosine-motifs (Morris and Cooper, 2001; Mishra et al., 2002). Dab2 mediates low density lipoprotein receptor (LDLR) endocytosis and numb downregulates Notch1 by ubiquitination through itch, an E3 ubiquitin ligase binding to numb (Mishra et al., 2002; McGill and McGlade, 2003). Both numb and Dab2, are linked to the endocytosis machinery by interacting with AP-2 and Eps15. Dab2, in addition, binds to clathrin and is targeted to the plasma membrane through interaction with the phospholipid PtdIns(4,5)P$_2$ (Santolini et al., 2000; Mishra et al., 2002). These PTB containing proteins interact with both cargo and the endocytic machinery of AP-2 and clathrin, suggesting that they might mediate endocytosis for certain cargo and act as adaptor proteins.

1.2.1.2 Accessory proteins

Many accessory proteins interact with AP-2 and/or clathrin, which in turn interact with other accessory proteins, all of them involved in endocytosis (Slepnev and DeCamilli, 2000). All of the proteins involved in endocytosis, as well as their interactions are displayed in Figure 2. They are grouped by color-code according to their function in endocytosis. Other accessory proteins include EH (Eps15 homology) domain-containing proteins, ENTH (epsin N-terminal homology) domain-containing proteins, proteins involved in the release of the vesicle from the membrane and finally proteins required for clathrin dissociation after vesicle. The proteins containing EH domains as well as the ENTH domain-containing proteins are all involved in the formation of clathrin-coated vesicles at the plasma membrane budding (Santolini et al., 1999; De Camilli et al., 2002). All of them interact with at least one other binding partner also involved in vesicle formation. The importance of some of these proteins is shown by their depletion or overexpression that causes an inhibition of internalization of cargo.

There are a number of domains and motifs that are found in many of these accessory proteins for interaction among each other. These domains and motifs include the EH domain that interacts with the NPF sequence, both the ENTH domain and the PH (pleckstrin homology) domain which bind to PtdIns(4,5)P$_2$ and the SH3 (Src homology 3) domain which interacts with a proline-rich domain (PRD) (Grabs et al., 1997; Salcini et al., 1997; Ford et al., 2001). The accessory proteins mostly contain more than one of these domains or motifs, forming numerous interactions (see Figure 2).

EH domain-containing proteins

Eps15 and the highly related Eps15R (Eps15 Related) contain three EH domains, several UIM, a PRD and fifteen DPF sequences which interact with the α adaptin ear of AP-2 (Benmerah et al., 1996; de Beer et al., 1998; Torrisi et al., 1999). A number of proteins involved in endocytosis such as Dab2, numb, epsin1 and synaptojanin1, contain an NPF sequence which interacts with Eps15. In addition, the coiled-coil domain of Eps15 is required for homo-dimerization as well as for hetero-dimerization with Eps15R and intersectin (Sengar et al., 1999). Overexpression of a dominant-negative form of Eps15 inhibited internalization of transferrin receptor and EGFR indicating that Eps15 is required for efficient receptor-mediated endocytosis (Benmerah et al., 1998).

Furthermore, Eps15 also binds to the γ adaptin ear of AP-1 and to Hrs indicating that Eps15 could also be involved in vesicle budding from the TGN and formation of multi-vesicular bodies in endosomes in addition to its function in endocytosis (Bean et al., 2000; Kent et al., 2002).

Intersectin forms a heterodimer with Eps15 through the coiled-coil domain and interacts with epsin1 through the EH domain. In addition, intersectin interacts with dynamin and synaptojanin, both required in the later stages of vesicle formation (Yamabhai et al., 1998; Sengar et al., 1999). Intersectin is suggested to function as a scaffolding protein in the organization of other endocytic regulatory components.

ENTH domain-containing proteins

The ENTH domain-containing proteins include epsin1, AP180/CALM, HIP1 and HIP1R (HIP1 related). The ENTH domain recruits these proteins to the plasma membrane.

Epsin 1 contains UIMs interacts with the α ear and the β2 ear of AP-2. The NPF motif of epsin1 was shown to interact with Eps15 and intersectin (Chen et al., 1998; Rosenthal et al., 1999). It has been shown that epsin1 promotes assembly of clathrin *in vitro* (Kalthoff et al., 2002).

AP-180 is a brain-specific protein and CALM is the ubiquitously expressed functional homologue of AP-180 in non-neuronal cells (Tebar et al., 1999). AP-180/CALM contains binding sites for clathrin and for the α ear as well as for the β2 ear of AP-2. In vitro experiments revealed that it can also promote assembly of clathrin (Hao et al., 1999).

HIP1 interacts with AP-2 and clathrin and forms heterodimers with HIP1R, which in addition has an actin binding site (Metzler et al., 2001; Waelter et al., 2001; Legendre-Guillemin et al.,

1.2.1.3 Proteins involved in vesicle 'pinching off'

Synaptojanin 1

The long isoform of synaptojanin1 is ubiquitously expressed and harbors AP-2 binding domains which the short, neuronal isoform lacks. Synaptojanin1 contains two DP[F,W] sequences, a FXDXF sequence and the recently discovered WXX[FW] motif, all of which interact with the α adaptin ear of AP-2 (Jha et al., 2003). Synaptojanin1 further consists of two catalytic phosphatase domains, the Sac1 (suppressor of actin 1) and the inositol-5' phosphatase homology domains, followed by a proline-rich domain (PRD) and an NPF sequence. Synaptojanin1 interacts with the EH domain of Eps15 and binds to SH3 domains of intersectin and amphiphysin2 (McPherson et al., 1996). The inositol-5' phosphatase dephosphorylates $PtdIns(4,5)P_2$ to $PtdIns(4)P$, which subsequently gets further dephosphorylated by Sac1 to phosphatidylinositol. Thus, synaptojanin1 is a phosphoinositol polyphosphatase that can generate phosphatidylinositol from $PtdIns(4,5)P_2$. $PtdIns(4,5)P_2$ is primarily found at the plasma membrane and recruits proteins containing a PH or an ENTH domain to the plasma membrane (Stauffer et al., 1998; Guo et al., 1999). The hydrolysis of $PtdIns(4,5)P_2$ is required for the release of the $PtdIns(4,5)P_2$ binding factors, such as AP-180/CALM, espsin1, HIP1/HIP1R and dynamin from the budded vesicle. Neurons derived from synaptojanin1-deficient mice show a delay in vesicle release and an accumulation of deeply invaginated clathrin-coated buds. These mice die shortly after birth due to neurological defects(Cremona et al., 1999). Thus, synaptojanin1 is thought to play a role in the separation of the vesicle from the membrane.

Endophilin

Endophilin contains an SH3 domain, through which it interacts with dynamin, amphiphysin2 and synaptojanin1 (Ringstad et al., 1997). The N-terminal domain of endophilin contains lysophosphatidic acid acyl transferase activity and binds to lipids (Schmidt et al., 1999). Depletion of endophilin resulted in an arrest of the invagination reaction at the stage of shallow coated pits, indicating that endophilin functions in the formation of membrane curvature. This effect may be mediated either by its binding to lipids or by the induction of an asymmetry in bilayer geometry due to conversion of lysophosphatidic acid to phosphatidic acid (Schmidt et al., 1999).

Amphiphysin 2

The N-terminal domain of amphiphysin2 mediates the formation of dimers and harbors a lipid-binding site that mediates plasma membrane targeting (Ramjaun et al., 1999). Amphiphysin2 contains a clathrin box and DP[F/W] and FXDXF motifs to interact with the α subunit of AP-2 (McMahon et al., 1997). In addition, amphiphysin2 comprises a SH3 and a PRD domain that bind to dynamin, synaptojanin1 and endophilin (Wigge and McMahon, 1998). These characteristics suggest that amphiphysin2 may act as a multifunctional adaptor that cooperates in the recruitment of coat proteins to the membrane and in targeting of synaptojanin1, endophilin and dynamin to the coat.

Dynamin

Dynamin comprises a GTPase activity, a PH domain that binds to PtdIns(4,5)P_2, a GTPase effector domain (GED) and a PRD domain that interacts with amphiphysin2, endophilin and intersectin (Salim et al., 1996). The GED is involved in dynamin oligomerization and self-assembly. Dynamin has a key function in the fission of clathrin-coated vesicles where its GTPase activity is essential.

1.2.2 Acting together for vesicle formation

The characterization of AP-180/CALM, epsin1 and Eps15 revealed their involvement in the early stages of the formation of clathrin-coated vesicle (CCV) formation. However, only AP-180 is enriched in CCVs, whereas epsin1 and Eps15 are not. Eps15 was restricted to the rim of the budding coated pit (Tebar et al., 1996; Chen et al., 1998). From studies on these proteins, a model evolved for coated pit formation. Epsin1 is recruited to the plasma membrane by binding to PtdIns(4,5)P_2 where it induces membrane budding (Ford et al., 2002). Epsin1 and the recruited AP-180/CALM facilitate the additional recruitment of clathrin. Epsin1 and Eps15 are then involved in recruiting and clustering AP-2 and become subsequently displaced while AP-2 triggers the polymerization of clathrin (Cupers et al., 1998). Epsin1 further modulates the curvature at the edge of the forming pit and the vesicle grows upon further recruitment of AP-2 (with cargo) and clathrin. AP-180/CALM and AP-2 were shown to play important roles in the regulation of vesicle size (Tebar et al., 1999; Ford et al., 2001). Additional accessory proteins, such as HIP1-HIP1R are involved in early stages of clathrin-coated pit formation, probably by recruiting clathrin and AP-2 to the clathrin-coated pit zones, defined by the actin cytoskeleton (Engqvist-Goldstein et al., 2001).

Amphiphysin2 links the coat proteins with the proteins required for the detachment of a coated vesicle from the plasma membrane. It also recruits dynamin, endophilin and synaptojanin1 to the clathrin-coated pit. Dynamin self-assembles into rings and tubules *in vitro* and was shown to form rings around the neck of a budding clathrin-coated vesicle *in vivo* (Takei et al., 1996; Muhlberg et al., 1997). The GTPase effector domain of dynamin acts as a GTPase activating protein and therefore, self-assembly of dynamin stimulates its GTPase activity (Sever et al., 1999). A model for the function of dynamin suggests that dynamin is recruited to the coated pit and forms stacks of rings around the stalks of coated pits (Takei et al., 1996). Whether dynamin can trigger the separation of the vesicle from the membrane by its GTPase activity or whether interacting proteins play a role is not yet known. One possibility is that coordinated GTP hydrolysis triggers a concerted conformational change in dynamin that tightens the rings around the neck of invaginated pits to such an extent that fission can occur (Hinshaw and Schmid, 1995). Another possibility is that the GTPase activity of dynamin is used to regulate effectors that mediate the fission reaction. Both endophilin and synaptojanin1 are recruited to the budding vesicle through dynamin and/or amphiphysin2 and contain lipid modifying activities which makes them candidates for the actual fission reaction. Endophilin might modify the lipid composition of the neck by converting an inverted-cone-shaped lipid (LPA) to a cone-shaped lipid (PA) and thereby inducing a negative curvature to the cytoplasmic leaflet of the membrane (Schmidt et al., 1999). The stretching and tightening of the dynamin collar in combination with the negative curvature of the plasma membrane may cause the neck to collapse and may drive the fission reaction (Kozlov, 2001). It is therefore suggested that dynamin and endophilin act jointly to induce fission. Synaptojanin1, an inositol polyphosphatase, would then facilitate the release of dynamin and other PtdIns(4,5)P_2-binding proteins by dephosphorylation of PtdIns(4,5)P_2 (Guo et al., 1999).

After the vesicle is budded off, it is uncoated and the components of the coat are recycled. Two proteins were identified to be responsible for clathrin dissociation, the G-cyclin-associated kinase (GAK), also called auxlin2, and Hsc-70, a constitutively expressed chaperone of the heat shock protein family (Morgan et al., 2001). GAK interacts with clathrin and the α ear of AP-2, and in addition with Hsc-70 through the J-domain, thereby recruiting Hsc-70 to the CCV (Umeda et al., 2000). Furthermore, the interaction of GAK with Hsc-70 stimulates the ATPase activity of Hsc-70, which then enhances the release of clathrin from clathrin-coated vesicles (Ungewickell et al., 1995).

1.3 Sorting in the TGN

The TGN is a tubulovesicular compartment and a central protein sorting station where the decision whether a protein is destined for secretion by the constitutive pathway or the regulated pathway or alternatively transported to the endosomal compartment by a selective pathway is made.

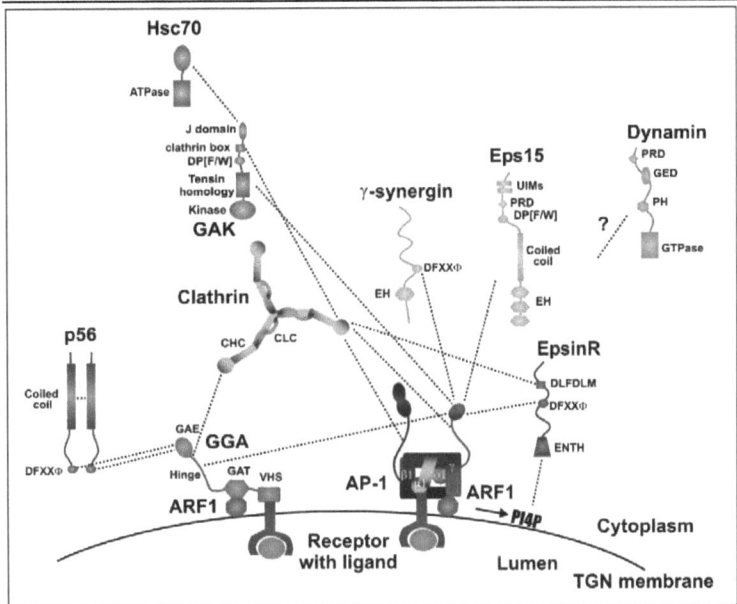

Figure 3: AP-1 and GGA with clathrin and accessory proteins. The domain structure of the proteins is not drawn to scale. The proteins are shown in different colours according to their function in vesicle formation and budding. The ENTH-containing proteins are in purple, EH-containing proteins are in green, proteins involved in the fission process from the membrane are in orange and proteins involved in the dissociation of clathrin after budding are in brown. The interactions are depicted by dotted lines. The names of the domains and the proteins are found in the text.

1.3.1 Transport from the TGN to endosomes

Proteins destined for endosomes or lysosomes are sorted away from the trafficking pathways followed by secreted proteins at the TGN and are instead targeted to the endocytic pathway compartments. Classical examples of proteins that follow this route are newly synthesized lysosomal hydrolases. These proteins obtain a mannose 6-phosphate tag on their N-linked oligosaccharides in the Golgi, representing the lysosomal targeting signal that is recognized by the MPR in the TGN (Kornfeld, 1992). The sorting in the TGN includes cargo such as the recycling membrane proteins (MPR, TGN38, sortilin and furin) as well as the lysosomal membrane proteins (LAMP1, LIMPII and CD63). All these proteins are destined for targeting to endosomes and onwards either to late endosomes and/or lysosomes, or alternatively to the plasma membrane.

Several adaptor proteins are recruited to the TGN membrane by ARF1, namely AP-1, AP-3, AP-4 and the GGAs, which are all involved in recruiting cargo for delivery to endosomes. AP-1 and GGAs require the clathrin coat, whereas AP-3 and AP-4 were suggested to function without clathrin.

1.3.1.1 Proteins involved in sorting in the TGN

ARF1

ADP-ribosylation factor 1 (ARF1) belongs to the group of small GTPases. The GDP-bound form of ARF1 is inactive and soluble, whereas the GTP-bound form binds tightly to the membrane (Goldberg, 1998). The membrane binding of ARF1 is regulated by a "myristoyl-ligand switch", where the myristoyl moiety is only exposed in the GTP-bound form of ARF1 (see also chapter: *N-Myristoylation*). The exchange of GDP with GTP of ARF1 is catalyzed by the guanine nucleotide exchange factor (GEF) of ARF1. ARF1 is recruited to the TGN and activated by a ARF GEF (Chardin et al., 1996). Activated, membrane-bound ARF1 recruits effectors such as AP-1, AP-3 and GGAs to the TGN (Stamnes and Rothman, 1993; Ooi et al., 1998a; Puertollano et al., 2001b). GTP hydrolysis of ARF1 is activated by a GTPase-activating protein (GAP), resulting in the release of ARF1 from the membrane (Donaldson, 2000). Brefeldin A (BFA), a fungal metabolite, inhibits the majority of the ARF GEFs, which subsequently blocks the activation cycle of ARF, leading to the disassembly of the Golgi and the block of secretion (Mansour et al., 1999). Thus, the BFA-sensitive targeting of a protein to a membrane indicates that an ARF is involved in the membrane recruitment of that protein.

Adaptor protein-1 (AP-1)

AP-1 is composed of the subunits γ1 or γ2, β1, µ1A or µ1B and σ1A or σ1B or σ1C (Robinson and Bonifacino, 2001). The AP-1 adaptins are ubiquitously expressed except for µ1B which is only expressed in polarized epithelial cells (Ohno et al., 1999). AP-1 is recruited to the TGN by binding of the trunk regions of γ adaptin and β1 adaptin to ARF1 (Stamnes and Rothman, 1993; Traub et al., 1993; Traub et al., 1995). In addition to the clathrin box in the β1 subunit hinge domain, there is a clathrin binding motif in the γ subunit hinge domain (Doray and Kornfeld, 2001). The AP-1 µ1 subunit binds to YXXΦ motifs in cargo proteins such as LAMP-1 and TGN38 and the µ1 and/or β1 subunit interacts with [D,E]XXXL[L,I] motifs in cargo proteins, such as the invariant chain (Ii) (Ohno et al., 1995; Höning et al., 1996; Ohno et al., 1996; Rapoport et al., 1998; Rodionov and Bakke, 1998). In addition, AP-1 interacts with phosphorylated casein kinase 2 sites in both, CD-MPR and CI-MPR (Le Borgne et al., 1993; Mauxion et al., 1996). The interaction of AP-1 with cargo is enhanced on phosphorylation of the µ1 subunit by the cyclin G-associated kinase (GAK), which binds to the γ ear of AP-1 (Umeda et al., 2000; Ghosh and Kornfeld, 2003a). The phosphorylation causes a conformational change that renders the cargo binding site in µ1 accessible. The γ ear of AP-1 interacts with the DFXXΦ motif in accessory proteins such as γ-synergin, Eps15 (both containing EH-domains) and epsinR (epsin-related, containing an ENTH-domain), which are thought to facilitate the formation of clathrin-coated vesicles (see Figure 3) (Page et al., 1999; Kent et al., 2002; Hirst et al., 2003). In addition, the AP-1 ear domain interacts with GGA, another adaptor protein described below (Doray et al., 2002b). The AP-1 complex is essential for viability since disruptions of the genes encoding γ1 or μ1A cause embryonic lethality in mice (Zizioli et al., 1999; Meyer et al., 2000).

For a long time AP-1 was thought to mediate the transport from TGN to endosomes (Ahle et al., 1988; Klumperman et al., 1993; Höning et al., 1997). But recent data about the function of AP-1 in sorting in the TGN are controversial. Fibroblasts deficient in μ1A shift the steady-state distribution of CD-MPR and CI-MPR to early endosomes and CD-MPR fails to recycle back from the endosomes to the TGN (Meyer et al., 2000). In addition, *in vitro* transport from endosomes to the TGN is impaired with membranes from μ1A-deficient fibroblasts and cannot be restored by adding cytosol containing AP-1 (Medigeshi and Schu, 2003). Furthermore, PACS-1 (phosphofurin acidic cluster sorting protein 1), a protein involved in transport from endosomes to the TGN, was shown to bind to AP-1 (Crump et al., 2001). Altogether, these results indicate that AP-1 or AP-1A (AP-1 complex containing the μ1A adaptin) is required for the endosomes to TGN transport of cargo. However, other models for the function of AP-1 involve AP-1 interaction with GGA in the TGN for proper sorting of cargo in the TGN (Doray et al., 2002b). One possibility could be the involvement of AP-1 in both anterograde and retrograde transport from the TGN to endosomes, wherein the specificity of transport would be provided by accessory proteins, such as PACS-1 and GGA. In contrast to the ubiquitously expressed AP-1 (AP-1A), AP-1B appears to be involved in basolateral sorting. The lack of μ1B expression in the polarized epithelial cell line LLC-PK1 caused impaired sorting of cargo to the basolateral plasma membrane surface (Folsch et al., 1999). AP-1A and AP-1B are localized to distinct subdomains in the TGN, suggesting that the differential sorting occurs in the TGN (Folsch et al., 2001).

γ-synergin

γ-synergin contains an EH (Eps15 homology) domain and binds to the ear domain of γ adaptin of AP-1 through DFXXΦ motifs (Page et al., 1999). γ-synergin is recruited to the TGN by binding to AP-1. It is enriched in CCVs and was therefore thought to be involved in the formation of AP-1 containing clathrin-coated vesicles at the TGN (Page et al., 1999). But its role in vesicle budding is not known.

EpsinR

EpsinR (epsin-related protein) contains an ENTH domain that interacts with PtdIns(4)P on the TGN membrane (Hirst et al., 2003; Mills et al., 2003). Membrane recruitment of epsinR is BFA-sensitive, indicating that ARF1 is required, possibly by stimulating the synthesis of PdtIns(4)P (Godi et al., 1999). EpsinR binds to the γ adaptin ear of AP-1 with the DFXXΦ motif and colocalizes with AP-1 at the TGN but membrane targeting of epsinR is AP-1 independent. Overexpression of epsinR disrupts lysosomal targeting of cathepsin D and incorporation of MPR into CCVs (Mills et al., 2003). However, depletion of epsinR by siRNA had no effect on cathepsin D trafficking (Hirst et al., 2003). It is suggested that epsinR is functionally equivalent to epsin1, but in CCV budding from the TGN/endosomes rather than from the plasma membrane (Mills et al., 2003). Recent studies revealed that the ENTH domain of EpsinR interacts with the v-SNARE vti1b (Chidambaram et al., 2003). This suggests that EpsinR might recruit the proper SNAREs to the clathrin-coated vesicle at the TGN. The yeast homologue, Vti1p, is involved in transport from TGN to endosomes (Fischer von Mollard and Stevens, 1999).

GGAs

In mammals, three GGAs were identified, GGA1, GGA2 and GGA3 (Boman et al., 2000; Dell'Angelica et al., 2000; Hirst et al., 2000). GGAs are monomeric cytosolic adaptor proteins, composed of four domains - an N-terminal VHS (Vps27p/Hrs/STAM) domain, a GAT (GGA and TOM1) domain, a connecting hinge segment and a C-terminal GAE (γ-adaptin ear) domain. Recruitment of GGA to the TGN is mediated by an interaction of the GAT domain with ARF1 (Collins et al., 2003), while the VHS domain binds to the DXXLL motif in cytosolic tails of cargo (Puertollano et al., 2001a; Takatsu et al., 2001; Misra et al., 2002). GGA interacts with clathrin through the clathrin box in the hinge domain of GGA, which binds in addition to the ear domain of the γ subunit of AP-1 (Puertollano et al., 2001a; Doray et al., 2002b). Furthermore, the GGA-GAE domain interacts with p56, a protein of unknown function. These interactions result in the formation of clathrin-coated vesicles emerging from the TGN (see Figure 3). GGA was shown to colocalize with clathrin and AP-1 in vesicles originating from the TGN and in some cases, GGA still localized to these vesicles when they fused with endosomes (Puertollano et al., 2003). The cargo of GGAs includes CI-MPR, CD-MPR, sortilin, memapsin 2 and LRP3 (LDL receptor related protein 3), albeit with different affinities to the three GGAs (Puertollano et al., 2001a; Takatsu et al., 2001; He et al., 2002). GGA is required for the sorting of cargo in the TGN and a dominant-negative form of GGA1 inhibits TGN exit of cargo such as CI-MPR and CD-MPR (Puertollano et al., 2001a). GGA1 and GGA3 binding to cargo is regulated by phosphorylation of an autoinhibitory site in the hinge domain of GGA1 and GGA3 (Doray et al., 2002a). This site is an internal DXXLL site that requires phosphorylation of a serine three residues upstream in order to bind to its own VHS domain, thereby inhibiting the interaction with cargo (Ghosh and Kornfeld, 2003b).

p56

p56, a protein of unknown function, was found to interact with the GGA-GAE domain (Lui et al., 2003). The N-terminal domain of p56 contains a DFXXΦ motif, known to interact with the γ ear domain of AP-1, which interacts with the GGA-GAE domain. Further domains of the p56 include a coiled-coil domain, involved in homo-dimerization and a short C-terminal domain of unknown function (Lui et al., 2003). p56 colocalizes with GGA at the TGN and is targeted to the membranes through binding to GGA and therefore membrane targeting of p56 is sensitive to BFA.

Adaptor protein-3 (AP-3)

AP-3 is composed of the following subunits, δ (homologues to γ and α), β3A or β3B, μ3A or μ3B and σ3A or σ3B. The two subunits, β3B and μ3B, are specifically expressed in neurons and endocrine cells, whereas all the others are ubiquitously expressed (Robinson and Bonifacino, 2001). AP-3 is localized to the TGN and endosomes and recruited to the membranes by interaction with ARF1 (Ooi et al., 1998b). AP-3 is required for targeting membrane proteins to lysosomes and lysosome-related organelles (Dell'Angelica et al., 1997; Dell'Angelica et al., 1999b). Although β3 adaptin can bind to clathrin, it is not required for the function of AP-3, which is therefore suggested to operate without clathrin (Simpson et al., 1997; Dell'Angelica et al., 1998; Peden et al., 2002). AP-3 binds to YXXΦ motifs in the cytosolic tails of membrane proteins with preferences for acidic residues surrounding the tyrosine, such as the lysosomal proteins CD63 and LAMP-1 (Ohno et al., 1996; Le Borgne et al., 1998; Rous et al., 2002). AP-3 also targets [D,E]XXXL[L,I] motif containing proteins to lysosomes and lysosome-related organelles, such as LIMPII and tyrosinase, a melanosomal protein (Höning et al., 1998; Le Borgne et al., 1998). Thus, AP-3 is suggested to transport lysosomal membrane proteins to lysosomes or lysosome-related organelles. Whether the transport occurs directly to lysosomes or via early or late endosomes is not known.

Recent data showed binding of AP-3 to PACS-1, a protein involved in transport from endosomes to the TGN and the requirement of cytosolic AP-3 in the *in vitro* transport from endosomes to the TGN (Crump et al., 2001; Medigeshi and Schu, 2003). These results indicate that AP-3 is also involved in the retrograde transport from endosomes to the TGN, in addition to the anterograde transport step.

The YXXΦ motif for lysosomal targeting is located 6 to 9 residues downstream of the trans-membrane domain (TMD) at the C-terminus of the lysosomal proteins. Changing the spacing of the GYQTI signal of LAMP-1 impairs targeting to lysosomes indicating that the placement of YXXΦ signals allows their recognition as lysosomal targeting signals at the TGN and/or endosomes (Rohrer et al., 1996). Whether this specificity is mediated by the AP-3 has not been investigated so far.

Adaptor protein-4 (AP-4)

AP-4 is composed of the subunits ε, β4, μ4 and σ4 (Dell'Angelica et al., 1999a; Hirst et al., 1999). AP-4 is localized to the TGN and its membrane association is BFA-sensitive, suggesting an interaction with an ARF protein for membrane targeting. AP-4 lacks a clathrin binding motif and is found in non-clathrin-coated vesicles, suggesting that AP-4 is part of a non-clathrin coat. *In vitro*, μ4 binds to YXXΦ motifs of TGN38 and the lysosomal proteins LAMP1, LAMP2 and CD63, albeit with a low affinity (Stephens and Banting, 1998; Aguilar et al., 2001). Depletion of μ4 caused a missorting of basolateral proteins to the apical surface, such as LDL receptor, CD-MPR and a Furin tail chimaera, indicating that AP-4 might be involved in basolateral sorting (Simmen et al., 2002). However, further investigations are required to determine the precise function of AP-4.

1.3.1.2 *Which cargo is selected by which adaptor protein*

All the proteins described above are suggested to be involved in the transport step from the TGN to endosomes. Some models evolved from the characterization of these proteins that suggest a subset of cargo and possibly distinct target organelles for the different adaptor proteins. Some of these adaptors proteins, AP-1 and AP-3 are suggested in addition to be involved in retrograde transport from endosomes to the TGN, possibly involving PACS-1.

Since the discovery of the GGAs, the function of AP-1 in the sorting of cargo in the TGN has become debatable. However, recent data suggests that the GGA and AP-1 cooperate in sorting in the TGN (see Figure 3) (Doray et al., 2002b; Puertollano et al., 2003). A model suggests that GGA, while recruiting cargo to the TGN, also binds to clathrin and AP-1 (see Figure 3). AP-1 in turn is associated with the CK2 and also interacts with clathrin. Once recruited to the complex CK2 then phosphorylates GGA and cargo, leading to the release of cargo from GGA due to autoinhibition and to an increase in affinity of AP-1 to cargo due to the phosphorylation. In this model GGA hands over cargo to AP-1 for further transport. However, GGAs were found to colocalize with clathrin, AP-1 and CD-MPR in vesicles leaving the TGN (Puertollano et al., 2003). This confirms on one hand the cooperation of the two adaptors, but on the other hand leaves the question, why GGA still remains associated with the vesicle, if it no longer binds to cargo. Thus, the mechanism of the cooperation between GGA and AP-1 requires further investigations. Nevertheless, the data indicate that GGA and AP-1 are both required for efficient sorting in the TGN. and might cooperate on a subset of cargo, the DXXLL-containing proteins. In addition to GGA, the accessory protein Eps15, might also be involved in recruiting and clustering AP-1 to prepared membrane buds, similar to its function at the plasma membrane (Tebar et al., 1996). Furthermore, EpsinR might be responsible for loading the budding clathrin-coated pit with the required v-SNARE (Chidambaram et al., 2003). The function of γ-synergin and p56 for this process remains to be investigated.

AP-1 was also shown to interact with both proteins containing a YXXΦ or a [D,E]XXXL[L,I] motif, suggesting that AP-1 might be responsible for the sorting of another set of cargo such as TGN38 and invariant chain (Ii), into CCVs in the absence of GGA. Whether AP-1 sorts this cargo into the same vesicles that contain the DXXLL motif containing cargo or whether AP-1 forms distinct vesicles, which are devoid of GGA, is not known.

The lysosomal proteins, LAMP-1, LIMPII and CD63 are probably selected as cargo by AP-3, although LAMP-1 was shown to bind to AP-1 *in vitro*. AP-3 is required for the exit out of the TGN and ultimately for lysosomal delivery of these lysosomal membrane proteins. Whether the pathway of the vesicles mediated by AP-3 includes a passage through endosomes or whether it is directly targeted from the TGN to lysosomes remains to be investigated.

Another subset of cargo is recruited by AP-1B, and probably AP-4, which are responsible in polarized cells for the sorting of proteins in the TGN destined for the basolateral plasma membrane.

Altogether, the TGN present a major sorting station, where cargo is sorted into different vesicles, clathrin-coated vesicles (AP-1 and GGA) and vesicles that are devoid of clathrin (AP-3 and AP-4). The different cargo is recruited to subdomains in the TGN through the interaction with adaptor proteins. However, the mechanism of the vesicle formation and precise information about the involvement of accessory proteins require further investigation.

1.3.1.3 'Pinching off' and clathrin dissociation

Dynamin was shown to be involved in the release of the CCVs from the TGN, in addition to its function at the plasma membrane (Jones et al., 1998; Kreitzer et al., 2000). However, the regulation of dynamin recruitment to the TGN and the identification of binding partners of dynamin require further investigation.

The GAK involved in uncoating CCVs derived from the plasma membrane (see above) also interacts with AP-1, in addition to AP-2 (Umeda et al., 2000). GAK and Hsc-70 are responsible for the uncoating of the CCVs similar to vesicles originating from the plasma membrane.

1.3.2 Sorting into secretory granules

The regulated secretory pathway, found in the more differentiated secretory cells is mediated by specialized secretory granules. Hormones and neuropeptides are secreted in this manner. The granin family (secretogranins/chromogranins) plays an important role in the sorting and aggregation of secretory products in the TGN as well as in the subsequent formation of secretory granules. Chromogranin B, also called secretogranin I, is a regulated secretory protein and member of the granin family and contains a disulfide-bonded loop at the N-terminus that acts as a signal for recruitment to the membrane of the TGN and delivery to secretory granules (Glombik et al., 1999).

Proteins, destined for regulated secretion, but lacking the loop, can be sorted via coaggregation with proteins containing the specific sorting signal (Gerdes and Glombik, 1999). Secretory granule formation apparently does not require a coat-driven budding process. Instead, it is thought that membrane deformation may result from the aggregation of secretory proteins in the TGN (Kim et al., 2001). Secretory granules bud from the TGN as immature secretory granules (ISG), containing proteins such as MPR and furin (an endoprotease), which are not destined for regulated secretion. Maturation of ISG includes removal of these proteins as well as homotypic fusion of ISGs. The homotypic fusions involve cytosolic components – NSF and α-SNAP. NSF and α-SNAP promote membrane fusion by priming SNAREs. The SNARE protein syntaxin 6 and possibly VAMP4 are involved in homotypic fusion of ISGs and are subsequently removed from the ISGs, since they are not detected on mature secretory granules (MSG) (Klumperman et al., 1998; Steegmaier et al., 1999; Wendler et al., 2001). Removal of furin, MPR with bound lysosomal enzymes, unprocessed secretory proteins, VAMP4 and syntaxin 6 from ISGs during maturation occurs in clathrin-coated vesicles (Dittie et al., 1997; Klumperman et al., 1998). Furin, VAMP4 and MPR recruit AP-1 to ISGs, wherein furin and VAMP4 require casein kinase 2 (CK2) dependent phosphorylation for AP-1 recruitment. The phosphorylated CK2 site in both furin and VAMP2, interact with PACS-1 which in turn binds to AP-1, and hence acts as a connector between the two proteins and AP-1. A dominant-negative PACS-1 mislocalizes furin and VAMP4 to mature secretory granules, confirming the requirement of PACS-1 in this trafficking step (for PACS-1 see also chapter: *Transport from endosomes to the* TGN) (Dittie et al., 1997; Hinners et al., 2003). AP-1 in turn recruits clathrin to the ISG and mediates clathrin-coated vesicle formation and budding. The vesicle is transported to and fuses with endosomes, where lysosomal proteins are delivered to the lysosome while unprocessed secretory proteins may be exported to the extracellular space, detected as constitutive-like secretion (Arvan and Castle, 1998). The contents of mature secretory granules (MSG) are released upon stimulation. The MSG fuses with the plasma membrane involving NSF and α-SNAP, which interact with a SNARE complex comprising VAMP/synaptobrevin (a v-SNARE), syntaxin1 and SNAP-25 (t-SNAREs) (Davis et al., 1999). Other proteins involved in regulated secretory granule exocytosis include Rab3A and its effectors, RIM and rabphilin 3A, as well as synaptotagmin, a Ca^{2+}-binding protein, which interacts with the SNAREs (Burgoyne and Morgan, 2003).

1.4 Sorting in endosomes

Cargo in the early endosomes is either recycled to the plasma membrane or transported to the TGN or alternatively delivered to multi-vesicular bodies (MVB)/late endosomes for lysosomal degradation. The lysosomal delivery is mediated by formation of vesicles, budded into the endosomes and forming MVB. For the transport of proteins from endosomes to the Golgi/TGN, several proteins were suggested to be involved, including the retromer complex, sorting nexins, PACS-1 and TIP47.

1.4.1 Formation of multi-vesicular bodies / late endosomes

A MVB was described by electron microscopy as an organelle consisting of a limiting membrane enclosing many internal vesicles (Marsh et al., 1986). MVBs are formed from early endosomes (EE) containing molecules that have been internalized and biosynthetic cargo from the TGN, including precursors of lysosomal enzymes. For the formation of MVBs two models have evolved. One model – the vesicular transport model – involves budding of large endocytic carrier vesicles from the EE, which subsequently form MVBs, followed by fusion with stable MVBs (Aniento et al., 1993). Alternatively, the MVB may represent an endpoint of a maturation process during which recycling components of EE are removed, a process which is called 'maturation model' (Futter et al., 1996). However, both models involve inward invagination occurring in EE or in a vesicle derived from EE, forming internal vesicles and resulting in MVBs. MVBs are also called late endosomes (LE) and ultimately fuse partially with lysosomes (Storrie and Desjardins, 1996).

The sorting of transmembrane proteins into topologically distinct limiting and intralumenal membranes has been proposed to serve several important functions (Raiborg et al., 2003): Firstly, transmembrane proteins in the intralumenal membranes are susceptible to degradation by lysosomal hydrolases, whereas proteins in the limiting membrane are resistant because only their lumenal region (which is usually protease-resistant due to extensive glycosylation) is exposed. Secondly, intralumenal vesicles might represent storage vesicles for transmembrane proteins that are destined to be released from the cell in a regulated manner as in secretory lysosomes (melanosomes, MHC II compartments and lytic granules). Thirdly, receptor signaling occurs from the limiting membrane of MVBs but not from the membranes of intralumenal vesicles.

Transmembrane proteins targeted for invagination into MVB are mono-ubiquitinated (Hicke, 2001). The mono-ubiquitination occurs at the plasma membrane and is carried out by E3 ubiquitin ligases that interact with cargo, such as EGFR, β_2AR, growth hormone receptor (GHR) and epithelial sodium channel (ENaC) (see Figure 4, upper section) (Raiborg et al., 2003). Many signaling receptors are mono-ubiquitinated in response to ligand binding. Although mono-ubiquitination takes place at the plasma membrane, ubiquitination has no effect on internalization of proteins but is required for cargo sorting in endosomes for delivery to lysosomes.

The following are two examples of receptors which undergo mono-ubiquitination, thereby obtaining the signal for lysosomal delivery. The β$_2$AR is phosphorylated in its cytoplasmic portion upon stimulation by the agonist isoproterenol, which enables interaction with arrestin3, leading to desensitization of the β$_2$AR. Arrestin3 interacts with Mdm2 and components of the endocytosis machinery (see Figure 4, upper section)(Shenoy et al., 2001; Laporte et al., 2002). Mdm2, the E3 ubiquitin ligase for p53, contains a catalytic RING finger that forms a thioester intermediate with ubiquitin through a conserved cysteine and transfers ubiquitin onto the substrate (Fang et al., 2000). Mdm2 mono-ubiquitinates arrestin3 and β$_2$AR which are subsequently internalized into endosomes. The E3 ubiquitin ligase for the EGFR is Cbl which also contains the characteristic RING finger and interacts directly with the phosphorylated tyrosine of the EGFR upon epidermal growth factor (EGF) stimulation (Waterman et al., 1999). The sorting for lysosomal degradation in the early endosomes is mediated by Hrs (hepatocyte growth factor-regulated tyrosine kinase substrate) (Raiborg et al., 2003). Depletion of Hrs inhibits the formation of MVB and leads to enlarged early endosomes and lysosomes, indicating that Hrs is a crucial component in the formation of MVB (Bache et al., 2003a).

Hrs was identified as a tyrosine-phosphorylated protein upon stimulation of cells with growth factor or cytokine (Komada and Kitamura, 1995). Hrs is composed of a VHS domain, a FYVE domain, an UIM, a coiled-coil domain and a clathrin box (Komada and Kitamura, 1995). The FYVE domain interacts with PtdIns(3)P, a lipid highly enriched on early endosomes and in the internal vesicles of MVBs and thereby targets Hrs to the membranes of early endosomes (Gaullier et al., 1998; Gillooly et al., 2000). Hrs, in turn, recruits clathrin which forms atypical flat lattices on the membranes of early endosomes (Raiborg et al., 2001a; Raiborg et al., 2002). Eps15, STAM1 (signal-transducing adaptor molecule) and STAM2 interact with Hrs forming a ternary complex (see Figure 4) (Bache et al., 2003b). STAM1 and STAM2 like Hrs become tyrosine-phosphorylated upon growth factor or cytokine stimulation of cells (Takeshita et al., 1996; Endo et al., 2000). Hrs, Eps15, STAM1 and STAM2, each contain an UIM and bind to ubiquitin and ubiquitinated cargo, albeit with a low affinity *in vitro*. Thus, the formation of the multi-UIM-containing complex might increase the affinity to ubiquitinated cargo (Raiborg et al., 2002; Mizuno et al., 2003). In addition, the UIMs act as a recognition signal for mono-ubiquitination, hence the four proteins undergo mono-ubiquitination upon stimulation by growth factor, with a possible function in stabilizing the formation of large complexes through multiple UIM-ubiquitin interactions (Polo et al., 2002). Mono-ubiquitinated cargo, destined for lysosomal degradation, colocalizes in subdomains on early endosomes with Hrs and the flat clathrin lattice, suggesting that the Hrs-Eps15-STAM1-STAM2 complex sorts mono-ubiquitinated cargo into subdomains on early endosomes, which are subsequently invaginated to form intralumenal vesicles leading to MVBs (Raiborg et al., 2003). The role of the flat clathrin lattice, which is not detected inside intralumenal vesicles, is not known. However, it was suggested to play a role in the endosomal retention of ubiquitinated membrane proteins before their inclusion into intralumenal vesicles (Sachse et al., 2002). For the actual inward invagination step, further protein complexes are involved – the ESCRT-I (endosomal complexes required for transport), ESCRT-II and ESCRT-III complexes (Babst et al., 2002). ESCRT-I is recruited to the early endosomes by Hrs through interaction of the PSAP motif in Hrs through its Tsg101 (tumor-susceptibility gene product 101) subunit (Bache et al., 2003a). ESCRT-I and ESCRT-II contain ubiquitin-binding subunits and act downstream of the Hrs-STAM1-STAM2-Eps15 complex. It is suggested that the ubiquitinated cargo can be relayed between these complexes. The subunits of ESCRT-III are small coiled-coil proteins that structurally resemble the SNARE proteins which mediate membrane docking and fusion. A possible scenario could be that ESCRT-III complexes, through stable coiled-coil interactions, function in inwards vesicle scission (Katzmann et al., 2002). Furthermore, the formation of MVB is also dependent on specific lipids. In addition to PtdIns(3)P located on limiting and intralumenal endosomal membranes which recruits Hrs and other FYVE domain containing proteins to endosomes, Lyso-bisphosphatidic acid (LBPA), a phospholipid with an inverted-cone shape, is a candidate for mediating inwards invagination of endosomal membranes. This lipid is required for MVB formation and found on intralumenal membranes (Kobayashi et al., 1998; Raiborg et al., 2001b). Fusion of the limiting (outer) membrane of the MVB with the lysosomal membrane results in the delivery of the lumenal MVB vesicles and their contents to the hydrolytic interior of the lysosome, where they are degraded (Futter et al., 1996).

1.4.2 Transport from endosomes to the TGN

1.4.2.1 The retromer complex

The retromer complex was identified to be essential for the transport of cargo, such as the Vps10p, the yeast functional equivalent of the MPR, from endosomes to the Golgi in yeast (Seaman et al., 1998). The retromer complex is composed of two subcomplexes, one is composed of Vps35p, Vps29p and Vps26p, which selects cargo for retrieval, and the second subcomplex contains Vps5p and Vps17p, which promotes vesicle formation. This complex has been conserved during evolution as human homologues of four out of the five components of the retromer complex were identified.

The human proteins, hVps35, hVps26, hVps29 and sorting nexin 1 (SNX1), the human homologue of Vps5p, form a complex in mammalian cells, that associates with membranes (Haft et al., 2000). Both, hVps26 and hVps29 bind to different sites in hVps35, the core protein of the complex, which, in addition, interacts directly with SNX1. The human homologue of Vps17p has not been identified so far. One component of the human retromer complex was shown to interact with other proteins; SNX1 binds to the EGFR and was suggested to be involved in the lysosomal degradation of the receptor (Kurten et al., 1996). However, SNX1 also interacts with Hrs through the same binding site as with EGFR (Chin et al., 2001). A possible scenario could be that SNX1 and Hrs compete for the substrate in early endosomes and Hrs regulates cargo recruitment by SNX1 by binding to the cargo binding site in SNX1 and therefore blocking its ability to recruit cargo (Chin et al., 2001). However, it is not obvious which transport step is mediated by SNX1, the transport to lysosomes, or possibly the transport from endosomes to the TGN, analogous to the retromer function in yeast.

1.4.2.2 Sorting nexins

SNX1 is part of a family of sorting nexins (SNX) that consists of a diverse group of cytoplasmic and membrane-associated proteins that are involved in various aspects of endocytosis and protein trafficking (Worby and Dixon, 2002). The common feature of the sorting nexins is a phox homology domain (PX), a sequence of 100-130 amino acids that binds to various phosphatidylinositol phosphates (PtdInsPs). The PX domains have a wide range of PtdInsP-binding specificities leading to the targeting of the SNX to distinct organelles that are enriched in some of the phospholipids (Ponting, 1996; Kanai et al., 2001). So far, 25 human SNX have been found which are divided into three subgroups. One subgroup (SNX1, SNX2, SNX4, SNX5, SNX6, SNX7, SNX8, SNX15, SNX16) consists of SNX that contain one to three coiled-coil domains in addition to the PX domain, possibly involved in homo- and/or hetero-oligomerization with other SNXs, as well as other protein-protein interactions. The second subgroup (SNX3, SNX10, SNX11, SNX12, SNX22, SNX23, SNX24) of SNX only contains the PX domain. The remaining SNX contain various protein-protein interaction sequences, such as a SH3 domain (SNX9, SNX18), a RGS (regulator of G-protein signaling) domain (SNX13, SNX14, SNX25) or other domains (SNX17, SNX19, SNX21, SNX27). Although for some SNXs the preference of lipids would direct them to the plasma membrane, a lot of these SNXs are localized to endosomes, indicating that isolated measurements of *in vitro* lipid binding affinities do not reflect the complex multiple interactions *in vivo*. Some sorting nexins were found to be involved in sorting steps in the endosomes, such as SNX1 (described above), SNX3, SNX13 and SNX15.

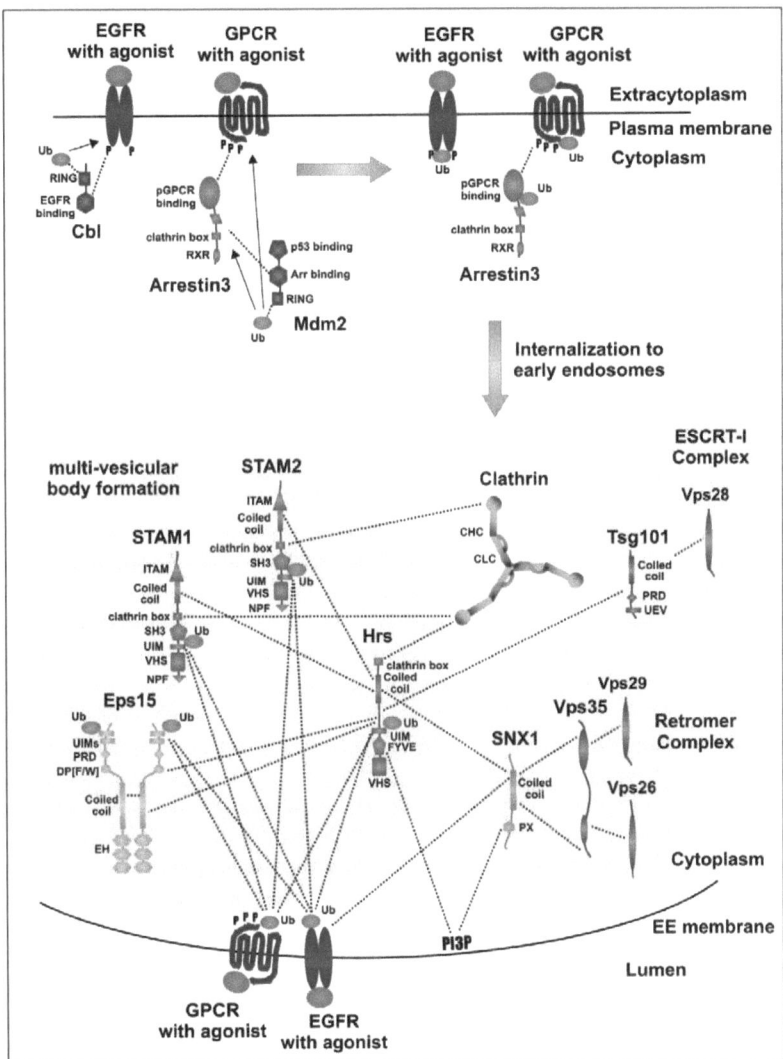

Figure 4 :Sorting in Endosomes. The upper section illustrates the mono-ubiquitination occurring at the plasma membrane by E3 ubiquitin ligases (purple). The sorting in endosomes is displayed in the lower section. The domain structure of the proteins is not drawn to scale. The proteins are shown in different colours according to their function in vesicle formation and budding. The adaptor and 'adaptor-like' proteins are in red, EH-containing proteins are in green, the PX-containing protein is in orange and other proteins are in blue. The interactions are depicted by dotted lines. The names of the domains and the proteins are found in the text. Abbreviation, which are not in the text: Ub, ubiquitin; UEV, ubiquitin conjugating enzyme E2 variant (an ubiquitin binding domain).

SNX3

SNX3 is recruited to early and recycling endosomes and colocalizes with EEA1 (early-endosomal autoantigen 1, a marker for early endosomes) and transferrin receptor (a marker for recycling endosomes) (Prekeris et al., 1998; Xu et al., 2001). Overexpression of SNX3 leads to an expansion of the tubulovesicular structure of the endosomes and a delay in EGFR degradation. Inhibition of SNX3 prevents transport of transferrin receptor from the early to the recycling endosomes. These results indicate that SNX3 has a function in membrane trafficking form early endosomes to recycling endosomes.

SNX13

SNX13 contains an RGS domain and functions as a GTP-activating protein (GAP) for the $G_{\alpha s}$-subunit of the heterotrimeric G-protein (Zheng et al., 2001). The PX domain localizes the SNX13 to early endosomes through interaction with PtdIns(3)P and colocalizes with EEA1. Overexpression of SNX13 inhibits degradation of the EGFR, suggesting a link between heterotrimeric G-protein signaling and protein sorting in the endosomes (Zheng et al., 2001).

SNX15

SNX15 localizes to early and late endosomes (Barr et al., 2000). Overexpression of SNX15 affects the morphology of endosomes, mislocalizes furin to endosomes and inhibits post-translational processing of insulin receptor (IR) and hepatocyte growth factor receptor (HGFR) precursors during their biosynthetic pathway (Phillips et al., 2001). Both IR and HGFR precursors are substrates of the endoprotease furin which is predominantly localized to the TGN, but cycles continuously from the TGN to the plasma membrane and endosomes, from where it is retrieved to the TGN through binding to PACS-1 (Komada et al., 1993; Wan et al., 1998). Thus, overexpression of SNX15 leads to a delayed processing of several furin substrates by mislocalizing furin, indicating that SNX15 is involved in the retrieval step of furin from the endosomes to the TGN (Phillips et al., 2001). Whether SNX15 and PACS-1 directly interact with each other is not known.

1.4.2.3 PACS-1

PACS-1 (phosphofurin acidic cluster sorting protein 1) binds to acidic clusters which contain a serine that is phosphorylated by casein kinase 2 (CK2). CK2 sites are composed of a serine surrounded by acidic residues between positions -4 and +7 with respect to the serine at position 0, the crucial residues being the acidic amino acids at position +1 and +3 from the serine (Meggio and Pinna, 2003). PACS-1 interacts with the acidic clusters of furin (a TGN-localized endoprotease) and VAMP4 (a SNARE protein on immature secretory granules) in a CK2 phosphorylation-dependent manner (Wan et al., 1998; Hinners et al., 2003). PACS-1 binds, in addition, to the acidic clusters of CI-MPR, however, independently of phosphorylation. Depletion of PACS-1 disrupts TGN localization of the proteins that interact with PACS-1 and leads to their subsequent mislocalization to mature secretory granules and/or endosomes, indicating that PACS-1 is essential for retrograde transport from endosomes and immature secretory granules back to the TGN (Wan et al., 1998; Hinners et al., 2003). Furthermore, PACS-1 binds to AP-1 and AP-3 and mediates the formation of a ternary complex between itself, AP-1 and membrane protein cargo, which suggests that PACS-1 functions as a connector by linking cargo to adaptor complexes (Crump et al., 2001; Hinners et al., 2003).

1.4.2.4 TIP47

TIP47 (MPR tail interacting protein of 47 kDa) interacts with CD-MPR and CI-MPR, albeit with different motifs in the two proteins (Diaz and Pfeffer, 1998). In the CD-MPR, TIP47 specifically recognizes the $F^{18}W^{19}$ motif, which has been identified to be essential for endosomal sorting of the receptor (Schweizer et al., 1997). The recognition of the CI-MPR by TIP47 is dependent on the three-dimensional structure and, in addition, requires the sequence PPAPRPG (residues 49-55) as well as the region from residues 55 to 75 in the cytoplasmic tail of the receptor (Orsel et al., 2000). Using membranes from CHO cells and cytosol from K562 cells (human), it was shown that depletion of TIP47 inhibits the *in vitro* transport of MPRs from late endosomes to the TGN, indicating that TIP47 is required for the retrograde transport of the receptors from late endosomes to the TGN (Diaz and Pfeffer, 1998). This transport step requires, in addition, Rab9 and is BFA-insensitive (Carroll et al., 2001). In mouse fibroblasts, however, the *in vitro* transport of CD-MPR from early endosomes to the TGN is TIP47-independent (Medigeshi and Schu, 2003). No other cargo was identified to bind to TIP47 so far and its physiological role has to be investigated further.

2 Mannose 6-phosphate receptor

Mannose 6-phosphate receptors (MPRs) recognize phosphorylated mannose residues, thus deriving their name. They are essential for the generation of functional lysosomes by directing newly synthesized soluble lysosomal enzymes bearing the mannose 6-phosphate signal to lysosomes. Two MPRs have been identified, the 46 kDa cation-dependent mannose 6-phosphate receptor (CD-MPR) and the 300 kDa cation-independent mannose 6-phosphate receptor (CI-MPR). Since the CI-MPR binds to insulin-like growth factor II (IGF-II) in addition to mannose 6-phosphate-containing ligands, it is also referred to as CI-MPR/IGF-II-receptor.

2.1 Structure and biosynthesis of MPRs

The two MPRs are type-I integral membrane proteins, containing an N-terminal signal sequence which is cleaved by the signal peptidase in the endoplasmic reticulum (ER). The mature protein comprises an extracytoplasmic ligand-binding domain, a transmembrane domain and a C-terminal cytoplasmic domain containing the signals for intracellular sorting. The amino acids of the extracytoplasmic regions are numbered starting from the N-terminus. However, to simplify the discussion of signals in the cytosolic domain these residues are numbered starting from the trans-membrane domain once again with residue number one.

2.1.1 The CD-MPR

The CD-MPR of 46 kDa is the smaller of the two MPRs. It consists of a 28-residue N-terminal signal sequence, a 159-residue lumenal domain, a 25-residue membrane span and a 67-amino acid cytoplasmic domain. The CD-MPR is a highly conserved protein with 93% overall homology among mammals, with a completely identical amino acid sequence within the cytoplasmic domain of human, mouse, pig and cow. CD-MPR homologues, have been identified in different species (for some species only partial sequences) - five mammals, one bird, two amphibians, and three bony fishes. The alignment of the complete sequences is shown in Figure 5.

2.1.1.1 Lumenal domain of the CD-MPR

After biosynthesis and insertion into the ER, newly synthesized CD-MPR undergoes signal peptide cleavage and N-linked glycosylation. Four out of five potential sites in the extracytoplasmic domain obtain N-linked oligosaccharides, two of them carry high mannose-type and the other two complex-type oligosaccharides (Wendland et al., 1991b). This glycosylation contributes a major

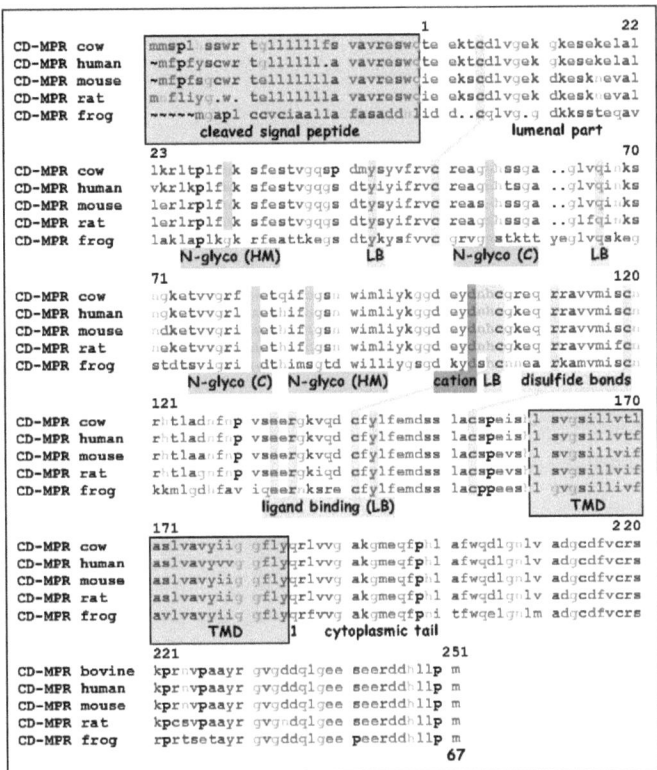

Figure 5: Alignment of CD-MPR from different species. The sequences are depicted in amino acid single letter code. The numbering starts from the N-terminus after signal peptide cleavage. The Asn acquiring complex type (C) or high mannose type (HM) oligosaccharides are marked in lila. The disulfide bonds are marked in blue. The residues involved in ligand binding (LB) are highlighted in yellow, wherein the cation-interacting Asp 103 is marked red. The TMD is indicated with a green box. The residues of the cytoplasmic tail are numbered starting from the TMD once again with residue number one (number below sequences).

part to the total mass of the CD-MPR (about 30%). However, glycosylation is neither involved in dimerization nor ligand-binding, intracellular stability or subcellular distribution of the CD-MPR (Wendland et al., 1991b; Marron-Terada et al., 1998a; Roberts et al., 1998). Instead, glycosylation promotes the proper folding of the receptor, although only one N-linked oligosaccharide is sufficient (Marron-Terada et al., 1998a). Another, even more important requirement for proper folding of the CD-MPR is the generation of three intramolecular disulfide bonds of six cysteine residues in its lumenal domain (Roberts et al., 1998). Mutation of a single cysteine disrupts the ligand-binding ability and impairs stability of the CD-MPR (Wendland et al., 1991a). Thus the conformational change caused by the formation of the disulfide bonds is required for ligand-binding ability. This is confirmed by the fact that the receptor acquires ligand-binding ability before entering the Golgi (Hille et al., 1990). Crystallization revealed that the extracytoplasmic domain alone is sufficient for dimerization and ligand-binding (Roberts et al., 1998). The CD-MPR is predominantly present as a dimer (67%) and also exists as a trimer or a tetramer and only a small portion is present in a monomeric state in the cell. The non-covalent dimerization is mediated primarily through hydrophobic interactions, comprising about 20% of the surface area of each CD-MPR monomer (Roberts et al., 1998). Dimerization and tetramerization increase the affinity of the receptor to the ligand. However, the oligomeric state is not altered for dissociation of the ligand from the receptor (Li et al., 1990). Furthermore, the oligomeric state is independent of pH and intracellular trafficking of the receptor (Punnonen et al., 1996; Olson et al., 2002).

Ligand binding of the CD-MPR depends on a divalent cation which specifically interacts with the aspartate 103 of the CD-MPR (Hoflack and Kornfeld, 1985; Tong and Kornfeld, 1989; Roberts et al., 1998; Olson et al., 1999). The relatively deep mannose 6-phophate binding pocket of the CD-MPR comprises the following residues: Tyr^{45}, Gln^{66}, Asp^{103}, Asn^{104}, His^{105}, Arg^{111}, Glu^{133}, Arg^{135} and Tyr^{143}, burying the terminal mannose and the phosphate group of the ligands (see c). Three-dimensional structures of ligand-free and ligand-bound receptor revealed that binding and dissociation from ligands is induced by a conformational change within the extracytoplasmic domain of the CD-MPR upon changes in pH (see a-b) (Olson et al., 2002). The CD-MPR binds ligands optimally at pH 6.3-6.5 (TGN), with a rapid decline as the pH approaches 5.5 (late endosomes) or 7.4 (plasma membrane) (Yamashiro and Maxfield, 1984; Tong et al., 1989; Machen et al., 2003). Upon pH change the release of the ligand is induced as well as the relocation of a loop (Glu^{134}-Cys^{141}) in the extracytoplasmic domain of the CD-MPR. The pK_α of Glu^{133} of the CD-MPR appears to be responsible for ligand release below pH5.5, whereas the pK_α of the sugar phosphate and His^{105} are accountable for the inability of the CD-MPR to bind ligand at the cell surface where the pH is about 7.4 (Stein et al., 1987; Olson et al., 2002). It is interesting to note that residues 102-105 are missing in the sequences of the two M6P recognition sites of the CI-MPR, which in contrast to the CD-MPR is cation independent and binds the M6P-ligand at the cell surface (see also chapter CI-MPR) (Hoflack et al., 1987).

2.1.1.2 Cytosolic tail of the CD-MPR

The CD-MPR cycles between the TGN, endosomes and the plasma membrane. The transport steps are mediated by sorting signals in the cytoplasmic tail of the CD-MPR, comprising 67 amino acids (see Figure 6). Given the absolutely identical sequences of the cytoplasmic tail in human, cow, mouse, and pig, every single residue and the three-dimensional structure of the cytoplasmic tail might play an important role in its trafficking.

Three internalization motifs

Three internalization sequences were identified in the cytoplasmic tail of the CD-MPR (Johnson et al., 1990; Denzer et al., 1997). The most potent of these motifs comprises the atypical Phe^{13}-X-X-X-X-Phe^{18} sequence with Phe^{18} being the key residue. The second signal involves Tyr^{45} as part of the typical tyrosine internalization motif, YXXΦ (where Φ is a bulky hydrophobic amino acid). The third sequence involves the dileucine motif at the C-terminus, L^{64}-L^{65}. For a maximal rate of receptor internalization all three motifs are required. Endocytosis is mediated by AP-2 that targets cargo to clathrin-coated pits. Interaction of AP-2 with the internalization motifs of the CD-MPR is controversial. While Höning (Höning et al., 1997) showed an interaction of AP-2 with the FXXXXF motif, but not with the tyrosine motif with surface plasmon resonance experiments, Storch (Storch and Braulke, 2001) demonstrated binding of AP-2 to the tyrosine motif, but not to the FXXXXF motif using yeast two-hybrid analysis. However, both reports show binding of AP-2 to the C-terminal region of the CD-MPR, dependent on the acidic cluster of the casein kinase 2 site, but independent of the di-leucine motif. The precise binding region(s) of the AP-2 to the CD-MPR *in vivo* and whether the CD-MPR requires additional interaction partners for internalization are not yet understood.

Basolateral sorting: Glu^{11} and Ala^{17}

In polarized cells, the portion of the CD-MPR that is localized to the cell surface is predominantly at the basolateral plasma membrane. The residues of the CD-MPR responsible for

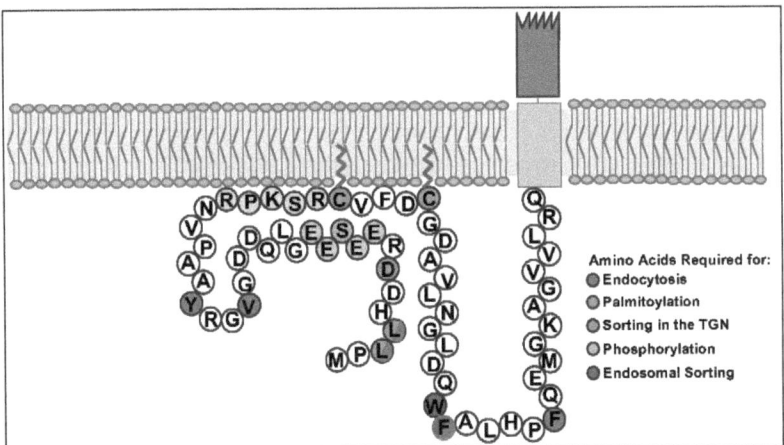

Figure 6: Model of the cytoplasmic tail of the CD-MPR. The sequence is displayed in the amino acid single letter code. The lumenal domain and the TMD are represented by boxes. The motifs involved in trafficking are shaded according to the legend. The residues involved in two trafficking steps are shaded in two colors.

basolateral sorting were revealed to be Glu^{11} and Ala^{17} (Bresciani et al., 1997). Single replacements of these amino acids resulted in missorting to the apical plasma membrane.

The diaromatic motif: Phe^{18}-Trp^{19}

The diaromatic motif Phe^{18}-Trp^{19} in the cytoplasmic tail of the CD-MPR is essential for the endosomal sorting of the receptor (Schweizer et al., 1997). When Phe^{18} and Trp^{19} are replaced by alanines the receptor is missorted to lysosomes. Changing the Phe^{18} and Trp^{19} to other aromatic amino acids was as efficient in CD-MPR sorting in endosomes as the wild-type receptor, while the replacement of Trp^{19} with hydrophobic residues, such as valine, isoleucine and leucine resulted in CD-MPR misrouting to lysosomes (Nair et al., 2003). Not only the aromatic residues but also their spacing from the TMD is crucial. Changing the spacing of the diaromatic motif from the TMD by adding five or deleting four amino acids led to an accumulation of the receptor in lysosomes (Schweizer et al., 1997). TIP47, a cytosolic protein, specifically binds to the diaromatic motif of the CD-MPR and is essential *in vitro* for the transport of the receptor from endosomes to the TGN, this transport step requiring the small GTPase Rab9 (Diaz and Pfeffer, 1998; Carroll et al., 2001). However, there are controversial reports on the requirement of TIP47 in the transport of the receptor from endosomes to the TGN (see also chapter: *TIP47*, page 35) (Medigeshi and Schu, 2003).

Palmitoylation of C^{30} and/or C^{34}

The CD-MPR is post-translationally modified by the reversible addition of palmitates to its cytoplasmic cysteines via a thioester bond (Schweizer et al., 1996). The palmitate rapidly turns over with a half-life of ~2 h, compared to a half-life of more than 40 h for the receptor. Mutation of either of the two cysteines still results in palmitoylation, which is abolished only when both cysteines, Cys^{30} and Cys^{34}, are replaced by alanines. The basic residues from position 35 to 39 (Arg^{35}, Lys^{37}, Arg^{39}) contribute to efficient palmitoylation. However, only the mutation of Cys^{34} leads to an accumulation of the receptor in lysosomes after incubation of the cells with lysosomal protease inhibitors (Pepstatin A and Leupeptin) for 24 h besides a total loss of CD-MPR mediated cathepsin D (a lysosomal enzyme) sorting in the TGN. Hence, the palmitoylation of Cys^{34} is essential for the proper trafficking of the CD-MPR from late endosomes back to the TGN (Schweizer et al., 1996). Palmitoylation of a cysteine residue 34 amino acids distal to the transmembrane domain (TMD) is rare among palmitoylated membrane proteins. Usually, palmitoylation of integral membrane proteins occurs within the TMD or in the range of 10 amino acids distal to the TMD. The palmitate residue is anchored to the membrane bilayer thereby inducing most likely, a drastic conformational change in the cytoplasmic tail of the CD-MPR. The fact that two prerequisites, the palmitoylation of Cys^{34} and the diaromatic motif exist for lysosomal

The phosphorylation of Ser^{57} and the CK2 site

CD-MPR is phosphorylated at Ser^{57} by casein kinase 2 (CK2) (Hemer et al., 1993; Körner et al., 1994). Stoichiometric analysis revealed that 3% of CD-MPR is phosphorylated at steady-state (Breuer et al., 1997). The half-life of the receptor-associated phosphate is 1.4 h in contrast to the long half-life of the CD-MPR. The functional importance of the CK2 phosphorylation site is under dispute. It was reported that neither the glutamates (Glu^{55}, Glu^{56}, Glu^{58}, Glu^{59}) surrounding the Ser^{57} nor the Ser^{57} itself inhibit sorting of cathepsin D (Johnson and Kornfeld, 1992b; Breuer et al., 1997). However, Mauxion (Mauxion et al., 1996) reported that mutating the glutamates impaired sorting of cathepsin D, whereas mutating Ser^{57} showed normal sorting of lysosomal enzymes. The phosphorylation of Ser^{57} of the CD-MPR, however, was required for delivery of the CD-MPR to the plasma membrane, suggesting that phosphorylation controls a sorting step within the endosomal system (Breuer et al., 1997). Furthermore, CD-MPR phosphorylation by CK2 was shown to be required for AP-1 binding (Mauxion et al., 1996; Ghosh and Kornfeld, 2003a). However, Höning (Höning et al., 1997) showed high affinity binding of AP-1 to non-phosphorylated CD-MPR peptides spanning the residues 49-67. Altogether, the physiological function of the phosphorylation of Ser^{57} is yet to be elucidated.

GGA binding: the Asp^{61}-X-X-Leu^{64}-Leu^{65} motif

GGAs are soluble monomeric adaptors essential for sorting of the receptor at the TGN into clathrin-coated vesicles which are destined for endosomes. GGA1, and to a lesser extent GGA3, were shown to interact with the DXXLL motif of the CD-MPR (Puertollano et al., 2001a; Misra et al., 2002). A dominant-negative form of GGA1 blocks the exit of the CD-MPR from the TGN. The importance of the DXXLL motif in trafficking from the TGN was confirmed by previous studies that found impaired cathepsin D sorting upon replacement or deletion of the dileucine motif, which forms a part of the DXXLL motif (Johnson and Kornfeld, 1992b; Mauxion et al., 1996). These studies showed that both the recycling rate of the mutant receptor from endosomes to the TGN as well as AP-1 binding to the mutant receptor were normal, indicating that the defect was in the sorting in the TGN, but independent of AP-1.

However, a different study revealed that the dileucine motif of the CD-MPR is required for sorting in the endosomes. The CD-MPR-$L^{64,65}$-A mutant accumulated in early endosomes and its transport to the TGN was impaired (Tikkanen et al., 2000). Thus, the dileucine motif seems to be required for retrograde transport from early endosomes to the TGN in addition to its function in sorting in the TGN.

AP-1 binding to residues 27 – 43 and 49 – 67

AP-1 was shown to bind to two distinct sites in the cytoplasmic tail of the CD-MPR (Höning et al., 1997). One site comprises the residues 27 to 43 including the cysteines (Cys^{30}, Cys^{34}) that are palmitoylated *in vivo* and the basic residues from 35 to 39 (Arg^{35}, Lys^{37}, Arg^{39}) that contribute to efficient palmitoylation (Schweizer et al., 1996). However, the CD-MPR peptides used in the AP-1 binding assay were synthetically produced and lacked palmitoylation. Thus, the effect of palmitoylation of the CD-MPR on binding to AP-1 was not analyzed. The second AP-1 binding site in the CD-MPR comprised the amino acids 49 to 67 of the cytoplasmic tail, including the acidic cluster of the casein kinase 2 site and the DXXLL that interacts with GGA (Höning et al., 1997). The specific binding site of AP-1 within the residues 49-67 was not identified. However, the dileucine motif (Leu^{64}, Leu^{65}) and the phosphorylation of Ser^{57} were excluded, since the binding affinity of AP-1 to the CD-MPR peptide was unaltered when the leucines were replaced by alanines and the peptides were not phosphorylated. In contrast, other studies revealed an increased AP-1 binding to the CD-MPR, when the Ser^{57} was phosphorylated by CK2 (Mauxion et al., 1996; Ghosh and Kornfeld, 2003a). Thus, the AP-1 binding site in the CD-MPR has to be further investigated.

2.1.2 The CI-MPR

The ~300 kDa CI-MPR is composed of a 44-residue N-terminal signal sequence, a large 2269-residue extracytoplasmic region, a single 23-residue transmembrane region, and a 163-residue C-terminal cytoplasmic domain.

2.1.2.1 Lumenal domain of the CI-MPR

The much larger extracytoplasmic region of the CI-MPR is composed of 15 domains that are similar to each other and to the extracytoplasmic domain of the CD-MPR. They are similar in size with an average of 147 residues per domain, share 14-38% amino acid sequence identity and display a comparable cysteine distribution, indicating that they exhibit similar cysteine bond formations and three-dimensional structures (Lobel et al., 1988). The lumenal region of the CI-MPR comprises two distinct carbohydrate recognition sites and a single IGF-II binding site (Tong et al., 1988).

The CI-MPR contains 19 potential N-linked glycosylation sites distributed throughout its extracytoplasmic region. Although it is not known how many of these sites are indeed glycosylated, it has been demonstrated that the CI-MPR acquires predominantly complex-type oligosaccharides (Sahagian and Neufeld, 1983). Comparable to the CD-MPR, N-linked glycosylation is not required for the acquisition of M6P or IGF-II binding ability by the CI-MPR (Sahagian and Neufeld, 1983; Kiess et al., 1991). The disulfide bonding pattern of the CI-MPR was predicted based on similarities in distribution to the CD-MPR. A similar requirement of disulfide bond formation for the proper folding of the CD-MPR is expected for the CI-MPR as well. The CI-MPR is mainly present as a monomer, but also exists as a weak dimer, which is stabilized by the binding of multivalent ligands. Furthermore, dimerization of the CI-MPR increases its rate of internalization at the cell surface (York et al., 1999).

The domains 3 and 9 of the extracytoplasmic region of the CI-MPR are the M6P-binding domains which have different binding affinities towards ligands and pH optima in ligand binding (Dahms et al., 1993; Marron-Terada et al., 1998b). The pH optimum for M6P binding of the domain 3 is pH 6.9, whereas the domain 9 binds M6P optimally at pH 6.4-6.5 (Marron-Terada et al., 2000). This reflects the broad pH range of the ligand binding exhibited by the CI-MPR including the binding of lysosomal enzymes at the cell surface at pH 7.4, a feature that is absent in the CD-MPR. The essential residues for M6P binding were revealed to be Gln^{392}, Ser^{431}, Arg^{435}, Glu^{460} and Tyr^{465} in the domain 3 and Gln^{1292}, His^{1329}, Arg^{1334}, Glu^{1354} and Tyr^{1360} in the domain 9 (Hancock et al., 2002). These results indicate that the two M6P binding sites in the CI-MPR and the binding site in the CD-MPR contain similar essential residues, forming the M6P binding pocket (Olson et al., 2002). However, the Asp^{103} and the His^{105} of the CD-MPR, involved in cooperation of the cation and in failure to bind ligands at the cell surface, respectively, are not found in the M6P binding domains of the CI-MPR. This reflects the cation-independent binding behavior of the CI-MPR as well as its ability to bind and internalize lysosomal enzymes at the cell surface (Hoflack et al., 1987).

The CI-MPR of viviparous mammals binds also the non-glycosylated polypeptide hormone, IGF-II (insulin-like growth factor II), at a binding site localized to the amino-terminal portion of the 11th extracytoplasmic domain. CI-MPR from marsupials, such as kangaroo and opossum, exhibit low affinities to IGF-II, whereas platypus, chicken and frog were unable to interact with IGF-II (Dahms and Hancock, 2002). In species that bind IGF-II, the I^{1572} of the domain 11 was determined as the key residue with a fibronectin type II domain within the domain 13 contributing to the IGF-II binding (Garmroudi et al., 1996; Devi et al., 1998).

2.1.2.2 Cytoplasmic tail of the CI-MPR

The cytoplasmic tail of 163 amino acids of the CI-MPR is longer than that of the CD-MPR of 67 residues. Like the CD-MPR, the cytoplasmic tail of the CI-MPR contains the sorting signals mediating the intracellular trafficking between organelles. The sequence is neither highly conserved among different species nor is it similar to the CD-MPR, however, corresponding sorting signals are found in the CI-MPR.

Palmitoylation of Cys^{15} and/or Cys^{16}

The CI-MPR is post-translationally modified by the attachment of palmitate residues to Cys15 and/or Cys16 of the cytoplasmic tail (Westcott and Rome, 1988). Palmitoylation of CI-MPR occurs via a thioester linkage (Schweizer et al., 1996). However, no function of the CI-MPR palmitoylation has been found so far. Compared to the CD-MPR the palmitoylated cysteines are located much closer to the TMD.

Internalization motif: Tyr^{24}-X-Tyr^{26}-X-X-Val^{29}

CI-MPR contains one tyrosine-based internalization motif, 24-YKYSKV-29, which interacts with AP-2 (Glickman et al., 1989; Canfield et al., 1991). The Tyr^{26} and Val^{29} are the key determinants for internalization whereas Tyr^{24} and Lys^{28} exhibit a contributing effect.

TIP47 binding to Pro^{49}-Pro-Ala-Pro-Arg-Pro-Gly^{55}

TIP47 binding to the CI-MPR is conformation dependent and requires the sequence 49-PPAPRPG-55 as well as the region from residues 55 to 75 in the cytoplasmic tail of the receptor (Orsel et al., 2000). Binding of TIP47 to the CI-MPR competes with binding of AP-1 and AP-2 to the tyrosine-motif at residues 24-29. In contrast to the extended binding region in the CI-MPR, the TIP47 binding motif in the CD-MPR comprises a short diaromatic motif, indicating that requirement for TIP47 binding differs a lot in the two cargo proteins known so far, CI-MPR and CD-MPR (Diaz and Pfeffer, 1998). TIP47 depletion leads to a decrease in CI-MPR half-life from $\tau \geq 35$ h to $\tau \sim 14$ h, indicating that TIP47 is essential for the proper sorting of the CI-MPR from endosomes to the TGN (Diaz and Pfeffer, 1998).

Phosphorylation of Ser^{85} and Ser^{156}

Both serines residues are part of a casein kinase 2 (CK2) site and are phosphorylated by CK2 *in vitro* and *in vivo* (Meresse et al., 1990). Phosphorylation was suggested to take place at the Golgi/TGN or in clathrin-coated vesicles leaving the Golgi/TGN (Meresse and Hoflack, 1993). Some studies revealed altered subcellular distribution of the CI-MPR upon phosphorylation or dephosphorylation, however, mutation of the serine residues had no detectable effect upon the steady-state sorting of lysosomal enzymes by the CI-MPR, indicating that phosphorylation might modulate the kinetics of the movement of the receptor rather than the regulation of the transport steps (Braulke and Mieskes, 1992; Johnson and Kornfeld, 1992a).

Further characterization of the CK2 sites revealed that the aspartate 157 adjacent to serine 156 within the C-terminal CK2 site is part of the recognition motif (DXXLL) for the GGAs, mediating sorting in the TGN (Puertollano et al., 2001a). The phosphorylation of the serine 156 had a contributing effect by increasing the affinity of GGA1 and GGA3 to the CI-MPR peptide (residues 154-163) (Kato et al., 2002). The acidic cluster of the CK2 sites is also involved in the interaction of the CI-MPR with PACS-1, independent of phosphorylation. PACS-1 mediates the transport of the CI-MPR from endosomes to the TGN (Wan et al., 1998). Thus, the acidic clusters of the CK2 sites have an important role in CI-MPR trafficking. However, to elucidate the precise function of the phosphorylation of the serines, further investigation would be required.

GGA binding: the Asp^{157}-X-X-Leu^{160}-Leu^{161} motif

GGA1, GGA2 and GGA3 bind to the C-terminal DXXLL motif in the cytoplasmic tail of the CI-MPR at the TGN (Puertollano et al., 2001a). The phosphorylated Ser^{156} had a contributing effect on GGA binding by increasing the binding affinity of the VHS domains of GGA1 and GGA3 to the CI-MPR peptide (residues 154-163) *in vitro* (Kato et al., 2002). The interaction with GGA at the TGN is required for the sorting of the receptor in the TGN and the transport to endosomes (Puertollano et al., 2001a; Puertollano et al., 2001b). The importance of the DXXLL motif is further confirmed by reports that revealed missorting of lysosomal enzymes, when the CI-MPR was truncated, thus lacking the dileucines or alternatively when the CI-MPR was analyzed by alanine cluster mutagenesis (Johnson and Kornfeld, 1992a; Chen et al., 1997).

AP-1 Binding to residues 26 - 29, 39 - 44, 84 - 88 and 154 - 160

Four distinct binding sites for AP-1 were identified in the CI-MPR. One binding site contains the tyrosine-based signal, 26-YSKV-29, which in addition acts as internalization motif and interacts with AP-2 (Ghosh et al., 2003). The second AP-1 binding site consists of a "dileucine"-like motif, 39-ETEWLM-44, which is a motif that AP-1 was shown to bind in cargo (Rodionov and Bakke, 1998). The remaining two AP-1 binding sites include the two CK2 phosphorylation sites, 84-DSEDE-88 and 154-DDSDED-160. The binding affinity of AP-1 to the CK2 sites is enhanced on phosphorylation (Le Borgne et al., 1993).

2.1.3 Comparison between the CD-MPR and the CI-MPR

Although both receptors are ubiquitously expressed in adult cells and tissues, there are cell-type and tissue-specific differences in the steady-state concentration of the two receptors. During mouse embryogenesis up to day 15.5, a non-overlapping distribution of the mRNA transcripts was observed for the two receptors, suggesting a distinct role for each of the MPRs during fetal development in the mouse (Matzner et al., 1992). Furthermore, quantitative studies demonstrated that there is upto 8-fold difference in the concentration of the CD-MPR and CI-MPR among several adult human cell lines and tissues, indicating that the expression of each receptor is independently regulated (Wenk et al., 1991).

The subcellular distribution of the CD-MPR and the CI-MPR was similar, being localized to the same organelles, the TGN, endosomes, cytoplasmic vesicles and the plasma membrane. However, the relative distribution of the two receptors among those organelles differs and in addition varies between different cell types (Klumperman et al., 1993). Interestingly the receptors colocalize with each other and AP-1 in the same vesicles leaving the TGN in HepG2 and BHK cells. In early and late endosomes, however, the CI-MPR localized to the central part of the endosomes, whereas the CD-MPR rather localized to associated tubules and vesicles, destined for recycling to the TGN. This explains the enrichment of CI-MPR over CD-MPR in the endosomal compartment (Klumperman et al., 1993).

In contrast to the CD-MPR, the CI-MPR binds to ligands at the cell surface, including M6P-tagged lysosomal enzymes, as well as other proteins (Hoflack et al., 1987). This allows the CI-MPR to recapture missorted lysosomal enzymes delivering them to the lysosomes and additionally to target non-lysosomal enzymes for degradation.

2.2 Function and relevance of MPRs

2.2.1 Lysosomal biogenesis

Lysosomes are acidic organelles containing numerous hydrolytic enzymes and playing an essential role in the degradation of internalized and endogenous macromolecules (Kornfeld and Mellman, 1989). Lysosomal membranes comprise a set of highly glycosylated, lysosomal-associated membrane proteins (LAMPs), such as LAMP1, LAMP2, CD63/LAMP3 and LIMPII (lysosomal integral membrane protein II). The lysosomal lumen contains upto 50 different lysosomal enzymes (e.g. proteases, lipases, glycosidases) digesting and degrading macromolecules delivered to the lysosomes.

The lysosomal membrane proteins are targeted to lysosomes through dileucine-based or tyrosine-based signals in their cytoplasmic tails. LAMP1, CD63 and LIMPII interact with AP-3 in the TGN, mediating the lysosomal delivery of these proteins. The strong glycosylation extent stabilizes these proteins, preventing their degradation in the lysosomes (Kundra and Kornfeld, 1999). The function of these membrane proteins in lysosomal biogenesis is not known, but it is suggested that the highly glycosylated lumenal domains of these lysosomal membrane proteins might prevent self-digestion of lysosomes.

Lysosomal function depends on the proper delivery of newly synthesized soluble acid hydrolases to the lysosomes. The initial steps in the biosynthesis of soluble lysosomal enzymes are shared by secretory proteins. The signal peptide of the protein is recognized by the signal recognition particle, which targets the protein to the translocon on the rough endoplasmic reticulum (rER). The nascent protein is inserted into the lumen of the ER, the signal peptide is cleaved by the signal peptidase and the asparagine residues are glycosylated obtaining an oligosaccharide composed of two N-acetylglucosamine (GlcNAc) residues, nine mannose (Man) residues and three glucose residues. The oligosaccharide undergoes trimming in the ER, while the protein is folded, resulting in the removal of the three glucose residues and one mannose residue. The specific signal for the targeting of lysosomal enzymes is the mannose 6-phosphate (M6P) tag on the N-linked oligosaccharides, which is obtained by the sequential action of two enzymes. The first enzyme transfers an N-acetylglucosamine (GlcNAc) 1-phosphate to a mannose residue, whereas the second enzyme later exposes the M6P marker by removing the covering GlcNAc. The enzyme responsible for the first reaction is the UDP-GlcNAc:lysosomal enzyme GlcNAc-1-phosphotransferase (phosphotransferase), a hexameric 540 kDa complex consisting of two disulfide-linked homodimers of 166 and 51 kDa subunits and two non-covalently associated 56 kDa subunits (Bao et al., 1996a). The phosphotransferase is localized to the ERGIC and cis-Golgi and recognizes the lysosomal enzymes by one or more lysine residues on the surface of their three-dimensional structure (Tikkanen et al., 1997). The phosphotransferase exhibits a specificity for transferring GlcNAc-P to C-6-hydroxylgroups of Man α1,2 Man sequences on the N-linked oligosaccharides of the lysosomal enzymes forming a phosphodiester bond. The second enzyme, GlcNAc-1-phosphodiester α-N-acetylglucosaminidase, a 272 kDa complex of four identical 68 kDa glycoprotein subunits, is also known as "uncovering" enzyme (UCE), removes the covering GlcNAc from the mannose 6-phosphate recognition marker on lysosomal enzymes (Kornfeld et al., 1998). UCE is synthesized as an inactive proenzyme which is activated only when it reaches the destination organelle, the TGN, by the endoprotease furin (Do et al., 2002). This might prevent UCE acting on other substrates, such as the UDP-GlcNAc, which is a sugar donor for Golgi GlcNAc-transferases, during its trafficking itinerary through the Golgi. UCE is primarily localized to the TGN, where it uncovers the M6P tags of lysosomal enzymes, however, it cycles between the TGN and the plasma membrane (Rohrer and Kornfeld, 2001). The function of this trafficking pathway of the UCE is not known, but it is suggested that UCE might ensure the proper transport of lysosomal enzymes by continuing its function of uncovering M6P-tags in vesicles leaving the TGN. However, this would require the presence of the mannose 6-phosphate receptors in the same vesicles; this is not known and requires further investigation. The uncovered M6P tag of lysosomal enzymes acts as the recognition marker for the mannose 6-phophate receptors (MPRs), which bind the ligands in the TGN.

The lysosomal enzymes are recruited by the MPR to a specific subdomain in the TGN and the MPRs in turn recruit cytosolic adaptors, which mediate the formation of the clathrin-coated vesicles. The MPR cytoplasmic tails bind to AP-1 and to GGAs, both of which mediate clathrin-coated vesicle formation (see also chapter: *Transport from the TGN to endosomes*, page 21). Since the discovery of the GGAs the precise role of AP-1 in sorting cargo in the TGN is unclear. One model involves GGA, AP-1 and clathrin for the formation of clathrin-coated vesicles, wherein GGA recruits the MPR as well as clathrin to the TGN (Doray et al., 2002b). In addition, GGA recruits AP-1 and the associated CK2 to the clathrin-coated pits. CK2 then phosphorylates GGA and the MPR, thereby causing the autoinhibition of GGA and release of the MPR, as well as an increase in the binding affinity of MPR to AP-1 (Doray et al., 2002a; Ghosh and Kornfeld, 2003a). Thus, in this model GGA hands over the MPR to AP-1. Subsequently a clathrin-coated vesicle is formed, possibly involving accessory proteins, and budded. GGAs, AP-1 and clathrin were shown to colocalize on clathrin-coated vesicles leaving the TGN and in some cases, the coats were still covering the vesicle when it fused with early endosomes (Puertollano et al., 2003). Upon fusion with early endosomes, the MPRs and the lysosomal enzymes are further transported to the late endosomes. Currently there are two models for this transport step, the vesicular model and the maturation model (Mullins and Bonifacino, 2001; Piper and Luzio, 2001). The vesicular model suggests that an endosomal carrier vesicle, also described as the multi-vesicular body, transports materials from early to late endosomes. Whereas the maturation model predicts that early endosomes are formed by the homotypic fusion of vesicles derived from the plasma membrane, which eventually mature into late endosomes. Both models provide for an intermediate between early and late endosomes, however the difference lies in whether the intermediate is a specific transport vesicle budded from early endosomes or it is what remains after removal of recycling components from an early endosome (Mullins and Bonifacino, 2001).

In early and late endosomes, the lower pH induces a conformational change in the ligand binding domain of the MPR, resulting in the dissociation of the ligand from the receptor. The lysosomal enzymes are further transported to lysosomes, whereas the MPR is transported back to the TGN to mediate another round of transport or alternatively the receptor is delivered to the plasma membrane, where it is rapidly internalized. For the delivery of material from late endosomes to lysosomes several models including maturation, vesicular transport, "kiss and run", and a fusion model have been suggested. However, recent results favor either the fusion model or the "kiss and run". The fusion between a late endosome and a lysosome creates a hybrid organelle, in which digestion would take place and subsequently lysosomes would be re-formed from it (Luzio et al., 2000). In this fusion model the lysosomal enzymes would automatically be delivered to the lysosomes, where many of them are proteolytically activated. Alternatively, the "kiss and run" model suggests transient fusion and fission processes, where materials destined for lysosomal degradation, as well as lysosomal proteins are delivered to lysosomes (Storrie and Desjardins, 1996). In contrast, the MPR is transported from late endosomes back to the TGN. The long half-life of the MPRs and the low accumulation of 4% of CD-MPR in the lysosomes within 24 h indicate that the sorting of the MPRs in endosomes into vesicles for transport to the TGN is very efficient (Schweizer et al., 1996). Of the proteins and complexes mediating transport from endosomes to the TGN (see also chapter *Sorting in endosomes*, page 29), several were suggested to be involved in retrograde transport of the MPRs to the TGN, including PACS-1, AP-1, AP-3 and TIP47. TIP47 binds to both MPRs and is required for their transport from late endosomes to the TGN in K562 cells, but not in mouse fibroblasts (Diaz and Pfeffer, 1998; Medigeshi and Schu, 2003). PACS-1 binds to the CI-MPR and PACS-1 depletion causes the redistribution of the CI-MPR to endosomes, indicating that PACS-1 is essential for the transport of the CI-MPR from endosomes to the TGN, which in addition requires AP-1 and/or AP-3 (Wan et al., 1998; Crump et al., 2001). Consistent with this, both MPRs are redistributed to endosomes in AP-1 deficient mice, indicating that AP-1 is essential for MPR retrieval to the TGN (Meyer et al., 2000). Furthermore, an *in vitro* transport assay in mouse fibroblasts requires membrane associated AP-1, as well as cytosolic AP-3 for the transport of the CD-MPR from endosomes to the TGN (Medigeshi and Schu, 2003). The data on PACS-1, AP-1 and AP-3 fit into a single model. A possible scenario is the involvement of PACS-1 and TIP47 at different sites in endosomes, with PACS-1 transporting MPRs from early endosomes and TIP47 from late endosomes to the TGN. Whether other proteins such as the retromer or sorting nexins, interact with the MPRs as well is not known.

2.2.2 Role of CI-MPR at the cell surface

In contrast to the CD-MPR, the CI-MPR can bind and internalize ligands at the cell surface. This enables the CI-MPR to recapture M6P-modified lysosomal enzymes that are not properly sorted form the secretory pathway at the TGN and deliver them to the lysosomes (Hille-Rehfeld, 1995).

In addition to lysosomal enzymes, the CI-MPR interacts with and internalizes M6P- and non M6P-containing non-lysosomal proteins. Thereby, the CI-MPR is involved in facilitating activation of the growth inhibitor transforming growth factor-β (TGF-β), modulating circulating levels of the potent cytokine leukemia inhibitory factor (LIF) and targeting IGF-II for degradation (O'Dell and Day, 1998; Blanchard et al., 1999; Godar et al., 1999). This suggests that the CI-MPR might play an important role in tumor suppression, in addition to the lysosomal biogenesis.

2.2.3 Ligands of the MPRs

Both MPRs bind M6P-containing lysosomal enzymes to target them to the lysosomes. Fibroblasts lacking both MPRs exhibit a massive missorting of multiple lysosomal enzymes leading to a decreased level of intracellular lysosomal enzymes of less than 20% compared to fibroblasts expressing both MPRs (Ludwig et al., 1993; Ludwig et al., 1994; Pohlmann et al., 1995). This results in an accumulation of undigested material in the lysosomes, a phenotype, which is similar to the I-cell fibroblasts. Fibroblasts, lacking either of the two MPRs, are only partially impaired in sorting lysosomal enzymes. Thus, neither of the two MPRs can substitute completely for the other MPR with respect to the targeting of lysosomal enzymes not even if overexpressed (Kasper et al., 1996). Analysis of the missorted lysosomal enzymes revealed that both MPRs have distinct but overlapping affinities for lysosomal proteins, indicating that the two receptors may interact in vivo with different subgroups of hydrolases (Ludwig et al., 1994; Pohlmann et al., 1995).

In addition to lysosomal enzymes, the CI-MPR binds to M6P-containing proteins, such as TGF-β precursor and LIF. Furthermore it interacts with non M6P-containing proteins, such as IGF-II through the IGF-II binding domain (Dahms and Hancock, 2002).

2.2.4 MPR-independent lysosomal targeting

I-cell disease (ICD) is caused by elimination or severe reduction of the activity of the phosphotransferase, one of the enzymes required to generate the M6P-tag, resulting in missorting of lysosomal enzymes. However, certain tissues and cell types isolated from ICD patients, such as liver, spleen, kidney, brain, and B-lymphocytes exhibit normal cellular lysosomal enzyme levels, suggesting that a cell-type specific, MPR-independent mechanism of lysosomal enzyme targeting exists (Kornfeld and Mellman, 1989; Glickman and Kornfeld, 1993).

2.2.5 Lysosomal storage disorders

The majority of lysosomal storage disorders includes diseases that are caused due to a defective lysosomal hydrolase. In fact, there is a known disease for almost every lysosomal enzyme. Another category of lysosomal storage disorders includes diseases that are caused due to defective lysosomal biogenesis. The defective protein might either be responsible for the acquisition of the recognition tag of the lysosomal hydrolases or for the trafficking of the lysosomal proteins. Some lysosomal storage diseases are described below.

Inclusion-cell (I-cell) disease / mucolipidosis II

I-cell disease is caused by a deficiency of phosphotransferase, the first enzyme of two involved in the generation of the M6P tag on lysosomal enzymes (Kornfeld and Mellman, 1989). This leads to a failure in the recognition of the enzymes by the MPRs and subsequently, the secretion of the enzymes. Consequently, the affected cells accumulate functionally inefficient lysosomes loaded with undegraded material, resulting in dense inclusion, hence the name inclusion-cell disease. The clinical consequences include disproportionate dwarfism, coarse facial features, and retarded psychomotor development (Olkkonen and Ikonen, 2000).

In I-cell disease, MPRs are unable to transport lysosomal enzymes to lysosomes, due to the lack of the recognition tag. Thus, fibroblasts lacking both MPRs mimic the conditions of I-cell disease (Ludwig et al., 1994).

A related defect with milder symptoms, pseudo-Hurler polydystrophy, is caused by a less drastic impairment of the phosphotransferase (Olkkonen and Ikonen, 2000).

Hermansky-Pudlak syndrome type 2

Two mouse models for the human genetic disorder Hermansky-Pudlak syndrome (HPS) have mutations in the AP-3 δ and β3A subunits, respectively (Kantheti et al., 1998; Feng et al., 1999). Subsequently, two human HPS patients with mutations in the β3A gene were identified (Dell'Angelica et al., 1999b). Both the mice and the humans have defects in lysosomes and lysosome-related organelles, in particular the melanosomes and platelet dense bodies resulting in hypopigmentation and prolonged bleeding (Hermansky and Pudlak, 1959). AP-3 is responsible for the targeting of membrane proteins to the lysosomes and lysosome-related organelles and is therefore required for the biogenesis of lysosome-related organelles (Höning et al., 1998).

Chediak-Higashi syndrome

The symptoms of Chediak-Higashi syndrome include hypopigmentation of the skin, eyes and hair, prolonged bleeding times, recurrent infections and abnormal NK-cell function (Ward et al., 2002). The enzyme responsible was identified to be a soluble 429 kDa protein called lysosomal trafficking regulator (LYST), defects of which lead to disrupted function of lysosomes and lysosome-related organelles. However, the mechanism how LYST regulates the function of lysosomes is unknown (Ward et al., 2002).

Griscelli disease

There are two defective enzymes in Griscelli disease, Rab27a and myosin Va (Menasche et al., 2000). It is suggested that Rab27a is required to recruit myosin Va to melanosomes (Hume et al., 2002). Rab27a, in addition, is required for regulated secretion in cytotoxic T lymphocytes (Stinchcombe et al., 2001). The lack of Rab27a and myosin Va leads to an accumulation of melanosomes in the perinuclear area (instead of the periphery) in melanocytes and uncontrolled T-lymphocyte and macrophage activation in Griscelli disease patients.

Fabry disease

This lysosomal storage disease results from mutations in α-D-galactosidase, a lysosomal hydrolase, leading to a labile protein that is degraded (Ohshima et al., 1997). Accumulation of neutral glycosphingolipids that have terminal α-linked galactosyl moieties in vascular endothelial cells causes renal failure along with infarctions and strokes in patients with Fabry disease (Ohshima et al., 1997).

Infantile neuronal ceroid lipofuscinosis (INCL)

Defects in protein palmitoylthioesterase 1 (PPT1); a lysosomal enzyme, causes severe neurodegenerative disorder (see also chapter: *Palmitoylthioesterases*) (Vesa et al., 1995). Lipidated thioesters derived from acylated proteins accumulate in cells from patients with INCL. Recombinant PPT1 reverses the accumulation of lipid thioesters in INCL lymphoblasts when delivered to cells through uptake by the MPR (Lu et al., 1996).

3 Palmitoylation

3.1 Lipid modifications

The covalent attachment of lipid moieties is an essential modification found on many proteins. Lipid modification increases the hydrophobicity of proteins and contributes to their membrane association. Three major forms of lipid modifications have been recognized so far in eukaryotic systems (also used by viral proteins): the co-translational N-terminal myristoylation of cytosolic proteins, the post-translational C-terminal prenylation of cytoplasmic proteins, and the most common modification – the post-translational addition of palmitate to many integral and peripheral membrane proteins (see Figure 7) (Resh, 1999; Farazi et al., 2001). Whereas myristoylation and prenylation are stable, permanent lipid modifications, the thioester bond that links palmitate to protein is labile and reversible.

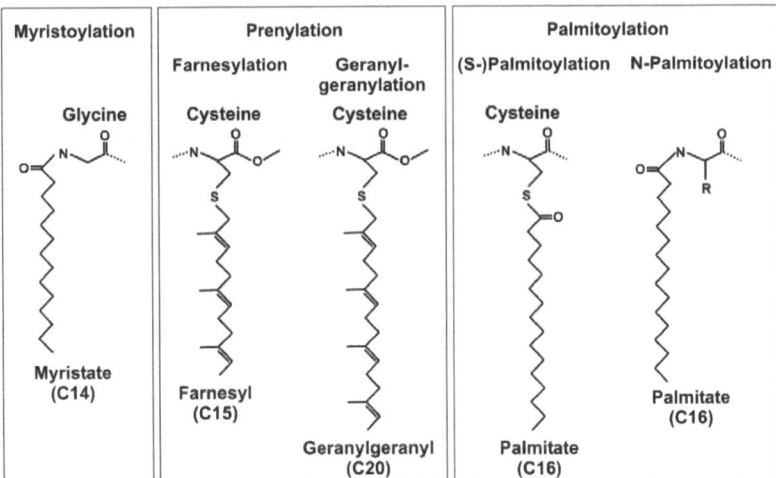

Figure 7: Lipid modifications. The structural formulas of the three different kinds of lipid modifications are displayed. The yellow box marks the amino acid to which the lipid modification is linked to. The number of carbon atoms of the lipid moiety is given in brackets.

3.1.1 N-Myristoylation

N-myristoylation refers to the covalent attachment of myristate, a 14-carbon saturated fatty acid, to the N-terminal glycine of eukaryotic and viral proteins (Resh, 1999; Farazi et al., 2001). It is an irreversible protein modification that occurs cotranslationally and is catalyzed by N-myristoyl transferase (NMT) (Towler et al., 1987). The consensus sequence for NMT protein substrates is:

Aim of the thesis

Met^1-Gly^2-X-X-X-Ser/Thr^6- and preferentially a basic amino acid at position 7 and/or 8. The initiating methionine is removed cotranslationally by methionine amino-peptidase and Gly^2 becomes the N-terminal amino acid. NMT catalyzes the transfer of myristate from myristoyl-CoA to the N-terminal Gly^2 linking it via an amide bond.

N-myristoylation promotes a weak and reversible protein-membrane interaction. For efficient membrane binding a second signal within the N-myristoylated protein is required. Some proteins, like several members of the src family of tyrosine kinases (Fyn, Lck, Yes) and certain α-subunits of heterotrimeric G proteins (G$_{\alpha i1}$, G$_{\alpha o}$) are dually fatty acylated with S-palmitoylation following N-myristoylation for efficient membrane anchoring (Milligan et al., 1995; Chen and Manning, 2001). They contain the consensus sequence Met^1-Gly^2-Cys^3- at the N-terminus with N-myristoylation of Gly^2 facilitating palmitoylation of Cys^3.

For other proteins like Src and MARCKS (myristoylated alanine-rich C kinase substrate), this second signal is a cluster basic of amino acids that binds to acidic headgroups of phospholipids in membranes (Sigal et al., 1994; McLaughlin and Aderem, 1995). Hence hydrophobic and electrostatic interactions act together to anchor the protein to the membrane.

The reversible membrane binding of some N-myristoylated proteins is regulated by throwing "myristoyl switches" between two conformations (Resh, 1999). In one conformation, the myristate moiety is hidden in a hydrophobic pocket within the protein. In the alternate conformation, myristate is flipped out and becomes available for membrane anchoring. The triggers for the myristoyl switch can be classified into three categories: electrostatics, ligand binding, and proteolysis. An example of the "myristoyl-electrostatic switch" is the MARCKS protein which is a protein kinase C (PKC) substrate that binds to membranes by its myristate and the basic motif as described above. Phosphorylation of MARCKS by PKC within the basic motif reduces the electrostatic interaction and results in the release of the protein from the membrane (McLaughlin and Aderem, 1995). The ARF proteins undergo "myristoyl-ligand switches" in which guanine nucleotide binding regulates the exposure of the myristate moiety. In the GDP-bound form, the myristoylated N-terminal helix is hidden in a hydrophobic groove within ARF1 (Haun et al., 1993; Amor et al., 1994). HIV-1 Gag is an example of a protein whose membrane binding is regulated by a "myristoyl-proteolytic switch". The HIV-1 Gag precursor, Pr55gag, binds to the plasma membrane via a myristate and a basic motif. Cleavage of Pr55gag by HIV-1 protease triggers a myristoyl-switch that results in the formation of the p17MA product, sequestration of myristate, and release of p17MA from the membrane (Hermida-Matsumoto and Resh, 1999).

3.1.2 Prenylation

Protein prenylation is the posttranslational attachment of either a farnesyl group or a geranylgeranyl group via a thioether linkage (-C-S-C-) to a cysteine at or near the carboxyl terminus of the protein (Maurer-Stroh et al., 2003; Roskoski, 2003). Farnesyl and geranylgeranyl groups consist of three and four isoprenes, respectively. There are three different protein prenyltransferases in humans: farnesyltransferase (FT) and geranylgeranyltransferase 1 and 2 (GGT1, GGT2), the substrate either being farnesyl-pyrophosphate or geranylgeranyl-pyrophosphate. Each protein consists of two subunits, α and β. The α-subunits of FT and GGT1 are the same, and the β-subunits differ, although all subunits are homologous to each other. All three prenyltransferases require Zn^{2+}, and FT and GGT2 additionally require Mg^{2+} for the transfer of the prenyl moiety from prenyl-pyrophosphate to substrates. FT and GGT1 catalyze the prenylation of substrates with a carboxy-terminal tetrapeptide sequence called a CaaX box, where "C" refers to a cysteine, "a" refers to an aliphatic residue, and "X" typically refers to methionine, serine, alanine, or glutamine for FT or to leucine for GGT1. FT and GGT1 are thus called CaaX prenyltransferases. Following prenylation of substrates, the terminal three residues (aaX) are subsequently removed by a CaaX endoprotease and the carboxyl group of the terminal cysteine is methyl esterified (Zhang and Casey, 1996). In contrast to FT and GGT1, GGT2 does not require a highly specific motif but recognizes the structural feature of the Rab-REP (Rab escort protein) heterodimer as substrate and is thus also called Rab geranylgeranyltransferase. The Rab proteins are geranylgeranylated at cysteines close to the carboxyl terminus that are often arranged as follows: -CC, -CXC, -CCX, -CCXX, -CCXXX or -CXXX. If the motif consists of two cysteines in close proximity, two geranylgeranyl moieties are usually added. Rab proteins ending with CXC residues are additionally methyl esterified while those ending with CC are not.

Many of the substrates of the prenyltransferases participate in signal transduction pathways related to cell growth, differentiation, cytoskeletal function, and vesicle trafficking. Typical substrates that are farnesylated by FT include many members of the Ras superfamily of small GTPases (H-Ras, K-Ras, N-Ras, RhoE). Some other small GTPases (Rac1, Rac2, RhoA) get geranylgeranylated by GGT1, as well as some γ-subunit variants of G proteins. The main substrates for prenylation by GGT2 are the Rab family of proteins, the largest group of small GTPases in the Ras superfamily. Some proteins (H-Ras, N-Ras) are additionally modified by palmitate following prenylation.

Aim of the thesis

3.1.3 Palmitoylation

Palmitoylation is a covalent attachment of a palmitate moiety to proteins. There are two kinds of palmitoylation: S-palmitoylation and N-palmitoylation.

3.1.3.1 N-Palmitoylation

N-palmitoylation, the rarer of the two kinds, occurs via an unusual amide-linkage of a palmitate to the amino group of the N-terminal amino acid. N-palmitoylation was first described for sonic hedgehog, a secreted signaling protein (Pepinsky et al., 1998). The N-terminal cysteine residue of hedgehog is modified by amide-linked palmitate. Skinny hedgehog (ski), which is also known as sightless, is required for palmitoylation of sonic hedgehog and shows a short but significant sequence homology to a diverse superfamily of membrane-associated acyltransferases (Chamoun et al., 2001). The fact that this family of enzymes catalyzes O-linked acylation transfers, strongly argues for a mechanism in which a thioester intermediate is formed with the side chain of the N-terminal cysteine, followed by a rearrangement through a five-membered cyclic intermediate to form the amide. Recently, N-palmitoylation has also been reported for $G_{\alpha s}$, the α-subunit of the heterotrimeric G-protein that activates adenyl cyclase (Kleuss and Krause, 2003). $G_{\alpha s}$ is modified at the N-terminal glycine (Gly^2 after removal of the starter methionine) by amide-linked palmitate and at the Cys^3 by thioester-linked palmitate. Cys^3-palmitoylation is reversible while Gly^2-palmitoylation is permanent and stable.

3.1.3.2 S-Palmitoylation

S-palmitoylation is the more common addition of a palmitate to proteins at cysteine residues via a thioester linkage. One of the specificities of the S-palmitoylation (which is simply called palmitoylation in this thesis) is its reversibility, which plays an important role in the signaling and trafficking of palmitoylated proteins (see also chapter: *Role of palmitoylation*, page 61).

3.1.4 Dual fatty acylation

N-myristoylation and prenylation are insufficient by themselves to stably anchor proteins to membranes. Typically, myristoylation and prenylation signals are linked to a second signal that assists in membrane anchoring. One secondary signal is a series of positively charged amino acids (e.g. in Src, K-Ras), another is palmitoylation. Among those proteins are members of the src family of tyrosine kinases (Fyn, Lck, Yes) and certain α-subunits of heterotrimeric G proteins ($G_{\alpha i1}$, $G_{\alpha o}$) which are first N-myristoylated, and GTPases (H-Ras, N-Ras) undergoing prenylation before palmitoylation. For all these proteins, myristoylation or prenylation is a prerequisite for subsequent palmitoylation (Hancock et al., 1989; Koegl et al., 1994). GAP-43 is an example of a cytosolic protein that is dually palmitoylated. Analysis of N-terminal sequences of dual fatty acylated proteins (Fyn, Yes, Lck, Gαo), fused to GFP, revealed that the myristate alone targeted the GFP to the ER and endosomes but not to the plasma membrane, while the myristoylation in combination with palmitoylation targeted the GFP to the plasma membrane and endosomes (McCabe and Berthiaume, 1999). Localization at the plasma membrane was also obtained when the N-terminal sequence of Src, containing a polybasic domain linked to the myristoylated sequence, was fused to GFP. Thus, myristoylation requires a second signal for localizing to the plasma membrane. The di-palmitoylated N-terminal sequence of GAP-43 targeted the GFP to the Golgi and the plasma membrane. Thus, di-palmitoylation confers a different specificity for membrane targeting from palmitoylation in combination with myristoylation. Altogether, the short dual fatty acylated sequence is sufficient for membrane targeting.

3.2 Palmitoylation motif of proteins

The types of proteins that undergo palmitoylation are diverse and include integral as well as peripheral membrane proteins. The palmitoylated proteins can be divided into four categories, group I comprises integral membrane proteins, group II includes proteins that are prenylated at the C-terminus, group III consists of proteins that are myristoylated at the N-terminus and group IV is made up of proteins that are only palmitoylated (see Table 1).

Most of the proteins of the group I are palmitoylated on cysteines within ten residues on either side of the trans-membrane domain (TMD) / cytoplasmic domain boundary. The composition of the TMD can influence the palmitoylation of the cysteines in the cytoplasmic domain. Replacing non-hydrophobic residues of the TMD of influenza virus hemagglutinin A (HA) with hydrophobic residues leads to a reduction in palmitoylation (Ponimaskin and Schmidt, 1998). Thus, the non-hydrophobic residues in a TMD can favor palmitoylation of membrane proteins. In some proteins, such as the G-protein-coupled β_2-adrenergic receptor (β_2AR), clusters of hydrophobic and positively charged amino acids around the palmitoylation site are required for palmitoylation or contribute to the reaction, as is the case for the CD-MPR (Schweizer et al., 1996; Belanger et al., 2001). In p63, the residues surrounding the palmitoylation site do not affect palmitoylation, however, the exact distance of seven amino acids from the TMD is crucial for palmitoylation of the cysteine (Schweizer et al., 1995). In contrast, some proteins are palmitoylated on cysteines that are further away from the TMD, such as the cysteine located 34 residues downstream of the TMD in the CD-MPR or 59 and 132 residues downstream of the TMD in the human immunodeficiency virus envelope glycoprotein 41 (Yang et al., 1995; Schweizer et al., 1996).

The proteins of group II, such as H-Ras and N-Ras, are farnesylated at their C-terminal CaaX sequence followed by palmitoylation of a cysteine in the C-terminal region. Prenylation is required for subsequent palmitoylation of the cysteines close to the prenylated cysteine (Hancock et al., 1989).

Group III consists of proteins that are cotranslationally N-myristoylated and undergo subsequent palmitoylation, such as Fyn, Lck, Yes, $G_{\alpha i1}$, $G_{\alpha o}$. In contrast to other palmitoylated proteins, these proteins contain a consensus sequence, Met^1-Gly^2-Cys^3-, at the N-terminus. Myristoylation of Gly^2 facilitates palmitoylation of Cys^3 and in some cases of another cysteine further downstream (Koegl et al., 1994).

The proteins of group IV are only palmitoylated, such as GAP-43 (growth associated protein 43), PSD95 (postsynaptic density protein 95), $G_{\alpha s}$, $G_{\alpha q}$, RGS4 (regulator of G-protein signaling 4) and SNAP-25b. For GAP-43 and PSD95, hydrophobic residues around the palmitoylation site are essential. $G_{\alpha s}$ is N-palmitoylated at the N-terminal glycine and S-palmitoylated at the adjacent cysteine. GAP-43, PSD95, $G_{s\alpha}$, $G_{\alpha q}$ and RGS4 are palmitoylated at the N-terminus, whereas SNAP-

Table 1: Palmitoylation Motifs	
Group I: integral membrane proteins	
p63	...SSSASCSRRLGR-*TMD*
LAT	*TMD*-CVRCRELPVSYDSTSTESLYPR
CD-MPR	*TMD*-QRLVVGAKGMEQFPHLAFWQDLGNLVADGCDFVCRSKPR...
CI-MPR	*TMD*-KKERREMVMSRLTNCCRRSANV...
Viral proteins:	
Influenza HA	*TMD*-CVKNGNMRCTICI
VSV G-protein	*TMD*-RVGIHLCIKLK...
GPCR:	
CCR5	*TMD*-EKFRNYLLVFFQKHIAKRFCKCCSIFQQ...
β₂-adrenergic receptor	*TMD*-SPDFRIAFQELLCLRRSSLK...
Endothelin B receptor	*TMD*-SKRFKNCFKSCLCCWCQSFEEK...
Group II: prenylated and palmitoylated proteins	
H-Ras	...SGPGCMSCKCVLS
N-Ras	...GTQGCMGLPCVVM
Group III: myristoylated and palmitoylated proteins	
Src family tyrosine kinases:	
Yes	M<u>G</u>CIKSKEDKGPAMKY
Fyn	M<u>G</u>CVQCKDKEATKLTE
Lck	M<u>G</u>CVCSSNPEDDWMEN
Gα subunits:	
αi1	M<u>G</u>CTLSAEDKAAVERS
αo	M<u>G</u>CTLSAEERAALERS
αz	M<u>G</u>CRQSSEEKEAARRS
Group IV: proteins modified exclusively by palmitoylation	
GAP-43	MLCCMRRTKQVEKNDDDQKIEQKGI...
PSD-95	MDCLCIVTTKKYRYQDEDTP...
RGS4	MCKGLAGLPASCLRSAKDMK...
Gα subunits:	
αs	M*****G***CLGNSKTEDQRNE...
αq	MTLESIMACCLSEEAKEA

Trans-membrane domains (TMD) are in italic letters followed by a hyphen. Palmitoylated cysteines are marked with bold letters. Prenylated cysteines or myristoylated glycines are underlined. A bold and italic letter marks the N-palmitoylated glycine.

Aim of the thesis

25b is palmitoylated at the C-terminus.

The diversity of palmitoylation sites and requirements for palmitoylation makes it difficult to predict whether specific proteins serve as substrates for palmitoylation. The lack of a clear consensus sequence and the diverse nature of the amino acids found to influence palmitoylation suggest that common structural features rather than strict sequence requirements are likely to be key factors specifying palmitoylation. Alternatively, several palmitoyltransferases might exist to account for the diverse requirements in structure and sequence of the various substrates. Some might recognize a specific amino acid environment of a cysteine (e.g. β_2AR) and others could be specific for a cysteine with a specific distance from the TMD, regardless of the neighboring amino acids (e.g. p63).

3.3 *Palmitoyltransferases*

Enzymatic as well as non-enzymatic palmitoylation has been described although the latter might only occur *in vitro*.

Non-enzymatic palmitoylation was reported for G_α subunits, rhodopsin and peptides of Yes (O'Brien et al., 1987; Duncan and Gilman, 1996; Bano et al., 1998). However, the non-enzymatic palmitoylation appear to be too slow to account for rapid palmitoylation of signaling proteins. In addition it would be difficult to achieve a tight regulation of the signal transduction pathways if the activation of key components would occur through a non-enzymatic process. Furthermore, the Acyl-CoA binding protein (ACBP) was found to inhibit specifically non-enzymatic palmitoylation, with almost no effect on enzymatic palmitoylation of G_α subunits (Leventis et al., 1997; Dunphy et al., 2000). Most of the cytosolic acyl-CoA in the cell is bound to acyl-CoA binding proteins (ACBPs). This indicates that under physiological condition there is probably only enzymatic palmitoylation.

The mechanisms involved in enzymatic palmitoylation are poorly understood. Some candidate enzymes have been found, although the characterization of the majority did raise doubts about their physiological relevance in palmitoylating their substrate or other proteins. A palmitoyltransferase activity was purified using mammalian H-Ras as a substrate but identified in the end as thiolase A. Its localization in peroxisomes makes it a very unlikely candidate for H-Ras palmitoylation (Liu et al., 1996; Liu et al., 1999). A 70 kDa palmitoyltransferase that adds palmitate to the cortical cytoskeletal protein spectrin has been isolated from erythrocytes, but no further characterization of this activity was reported (Das et al., 1997). A dimer of 260 and 270 kDa proteins which enhances palmitoylation of *Drosophila* Ras *in vitro* has been cloned from the silkworm *Bombyx mori* (Ueno and Suzuki, 1997). This protein complex is expressed only during embryogenesis and is probably not involved in palmitoylating Ras. In *Drosophila* skinny

Aim of the thesis

hedgehog/sightless was found to be required for palmitoylation of sonic hedgehog (Chamoun et al., 2001). Sonic hedgehog is modified by cholesterol at the C-terminus and palmitoylated through an atypical cysteine amide linkage at the N-terminus (Porter et al., 1996; Pepinsky et al., 1998). However, skinny hedgehog activity is on the luminal side of organelles of the secretory pathway and is therefore unlikely to directly palmitoylate cytosolic or transmembrane proteins on the cytosolic side. In yeast two palmitoyltransferases have been found, Akr1p that palmitoylates casein kinase Yck2p and the Erf2p/Erf4p complex that adds palmitate to Ras2p (Lobo et al., 2002; Roth et al., 2002). Erf2p and Akr1p are integral membrane proteins that both contain a conserved Asp-His-His-Cys cysteine-rich domain (DHHC-CRD), but share no other homology. Erf2p is localized to the ER and Akr1p, to the Golgi (Lobo et al., 2002; Roth et al., 2002).

Of all the described palmitoyltransferases, the two DHHC-CRD-containing proteins in yeast seem to be the most promising candidates. The DHHC-CRD is a zinc-finger binding motif and is also known as the NEW1 domain. Twelve human proteins, containing a NEW1 domain, were identified, but none of them have been cloned and tested for palmitoyltransferase activity so far.

3.4 Intracellular sites of palmitoylation

In addition to the above described palmitoyltransferases, many palmitoyltransferase activities have been reported mostly along the exocytic pathway. Some proteins (vesicular stomatitis virus G, influenza HA, CCR5) are palmitoylated in the ER-Golgi intermediate compartment (ERGIC) or cis-Golgi (Bonatti et al., 1989; Veit and Schmidt, 1993; Blanpain et al., 2001). p63, an ER-resident protein, is palmitoylated upon redistribution of the Golgi to the ER during mitosis or induced by a treatment with BFA (Mundy and Warren, 1992; Schweizer et al., 1993b). This indicates that p63 is palmitoylated by a palmitoyltransferase localized to the ERGIC or Golgi. Further in the secretory pathway, a palmitoyltransferase activity localized to the Golgi palmitoylates N-Ras (Gutierrez and Magee, 1991). For many cytosolic and membrane proteins (Fyn, G_α-subunits, transferrin receptor) palmitoylation seems to occur at the plasma membrane (Adam et al., 1984; Dunphy et al., 1996; van't Hof and Resh, 1997; Fishburn et al., 1999). Moreover, there is a palmitoyltransferase activity in mitochondria (Corvi et al., 2001; Veit et al., 2001) and in yeast on the vacuole (Veit et al., 2001). The various subcellular sites and the diversity of substrates for palmitoylation suggest that numerous palmitoyltransferases with different subcellular locations and specificities might exist.

3.5 Palmitoylthioesterases

The finding that the rate of palmitate turnover exceeds that of the protein itself for many palmitoylated substrates indicates that not only palmitoyltransferases but also protein palmitoylthioesterases are present in eukaryotic cells. Three such enzymes have been identified and

Aim of the thesis

characterized, the lysosomal hydrolases, protein palmitoylthioesterase 1 and 2 (PPT1, PPT2), and the cytoplasmic enzyme acylprotein thioesterase 1 (APT1) (Camp and Hofmann, 1993; Camp et al., 1994; Verkruyse and Hofmann, 1996; Soyombo and Hofmann, 1997b; Duncan and Gilman, 1998). They all share the conserved residues, a serine, an aspartate and a histidine, which are distant in the primary structure, but get arranged in proximity upon acquisition of the three-dimensional structure to form the active site (Devedjiev et al., 2000; Calero et al., 2003).

PPT1 and PPT2 are lysosomal hydrolases which are targeted to that organelle by modification of its oligosaccharides with mannose 6-phosphate (Verkruyse and Hofmann, 1996; Soyombo and Hofmann, 1997b). PPT1 deacylates cysteine thioesters in a variety of contexts, including intact proteins (palmitoylated H-Ras, Gα, and albumin), palmitoylated peptides, and palmitoyl-cysteine. PPT2, whose amino acid sequence is 28% identical to that of PPT1, hydrolyzes acyl-CoA (Soyombo and Hofmann, 1997b). But PPT2 is inactive with the other substrates of PPT1, suggesting that it has a role in degrading other types of lipid thioesters. APT1, initially purified as a lysophospholipase, cleaves thioesters in acyl-CoAs and acylproteins, as well as oxyesters in lysolipids (Duncan and Gilman, 1998). APT1 has been shown to depalmitoylate G_{i-1}, RGS4 (regulator of G protein signaling 4), and H-Ras *in vitro* with higher catalytic efficiencies than lysophosphocholine or palmitoyl-CoA (Yeh et al., 1999; Duncan and Gilman, 2002). The pronounced substrate preference for acylproteins, over lipid substrates, is consistent with a role for APT1 as a regulator of protein thioacylation and not as a regulator of lipid metabolism. More substrates have been reported to get deacylated by APT1 *in vitro* like many viral glycoproteins and endothelial nitric-oxide synthase (Yeh et al., 1999; Veit and Schmidt, 2001). Although viral glycoproteins do not undergo palmitate turnover in cells, this finding indicates that APT1 can also act on typical membrane proteins.

3.6 Role of palmitoylation

Many cytosolic proteins require palmitoylation for membrane association and in particular, for targeting to the plasma membrane and the Golgi. Furthermore, characterization of many palmitoylated proteins revealed an involvement of the palmitoylation in signal transduction, lipid rafts localization and in protein trafficking.

3.6.1 Palmitoylation in signal transduction

The signal transduction pathway through G-protein coupled receptors (GPCR) involves heterotetrameric G-proteins and additional proteins such as GPCR kinases (GRK) and regulators of G-protein signaling (RGS). Many members of these families of proteins undergo palmitoylation. Some proteins were characterized in more detail and a function of the palmitoylation was revealed.

Aim of the thesis

3.6.1.1 Palmitoylation of the G-protein

G-protein coupled receptors (GPCR) contain seven trans-membrane domains and transduce extracellular stimuli into intracellular signals via heterotrimeric GTP-binding proteins (G-proteins) (Qanbar and Bouvier, 2003). Each heterotrimer is composed of an α-subunit (G_α) and a βγ-subunit ($G_{\beta\gamma}$). There are many isoforms of these subunits in the cell and each receptor interacts with heterotrimers made up of distinct combinations of G-protein subunits. The G_α-subunit is palmitoylated, whereas the G_γ is prenylated. Palmitoylation of the G_α-subunit increases the association to $G_{\beta\gamma}$-subunits (Iiri et al., 1996). Consistent with this, $G_{\beta\gamma}$ protects the GDP-bound form of G_α, but not the GTPγS-bound form from depalmitoylation. The first step in GPCR activation is the binding of ligand to the receptor, which induces a conformational change in the receptor. This change, in turn, results in the engagement of the G-proteins. This interaction catalyzes the replacement of the GDP bound to the G_α-subunit with a GTP molecule. The heterotrimer then dissociates into G_α- and $G_{\beta\gamma}$-subunits which allows the depalmitoylation of the G_α-subunit. The simultaneous depalmitoylation of the receptor leads to the dissociation of the receptor-G_α interaction resulting in desensitization. The activated, depalmitoylated G_α then interacts with palmitoylated RGS proteins that inhibit G-protein signaling by promoting the intrinsic GTPase activity of G_α (Tu et al., 1997; Tu et al., 1999). GDP-bound G_α can then be repalmitoylated, which is a rapid process, since roughly the same proportion of all G_α is palmitoylated at any given time (Jones et al., 1997). Repalmitoylation facilitates coupling to $G_{\beta\gamma}$ forming the heterotrimer that is readily available for another round of activation.

3.6.1.2 Palmitoylation of the G-protein-coupled receptor

Many GPCRs, such as rhodopsin, β₂-adrenergic receptor, vasopressin 2 receptor, bradykinin B₂ receptor and CCR5 chemokine receptor, are palmitoylated on a cysteine residue proximal to the cytoplasmic end of the seventh trans-membrane domain, (Qanbar and Bouvier, 2003). For some GPCRs, such as the CCR5, palmitoylation occurs in the biosynthetic pathway and is required for targeting to the plasma membrane (Blanpain et al., 2001). For other GPCRs the lack of palmitoylation appears to have differential effects on the various signaling pathways engaged by a given receptor. For example, the unpalmitoylated human endothelin (ET)$_A$ receptor was less effective in stimulating $G_{\alpha i}$ and $G_{\alpha q}$, but as effective as wild-type in stimulating $G_{\alpha 0}$ (Doi et al., 1999).

β₂-adrenergic receptor (β₂AR), a hormone-binding G-protein coupled receptor (GPCR) is palmitoylated on a cysteine residue proximal to the cytosolic end of the seventh trans-membrane domain. Stimulation by the agonist isoproterenol induces depalmitoylation of the β₂AR, which is followed by a cascade of events (Moffett et al., 2001). Depalmitoylation of the receptor promotes

Aim of the thesis

phosphorylation by the cAMP-dependent protein kinase (PKA) of four serines, two of which are very close to the palmitoylation site and not accessible, when β_2AR is palmitoylated. Phosphorylation by PKA in turn facilitates phosphorylation of downstream sites in the C-terminal portion of the receptor by GPCR kinase 2 (GRK2) (Moffett et al., 2001). The phosphorylated β_2AR binds to arrestin 3 which in turn uncouples the receptor from the G-protein $G_{\alpha s}$ thereby leading to desensitization of the β_2AR. Arrestin 3 has many binding partners itself, such as clathrin, AP-2 and Mdm2 (a E3 ubiquitin ligase), that are all involved in directing the β_2AR to clathrin-coated pits and subsequent internalization (Shenoy et al., 2001; Laporte et al., 2002). In early endosomes, β_2AR is either targeted for degradation and sorted into internal vesicles by mono-ubiquitination or it is recycled to the plasma membrane and dephosphorylated. For recycling, dephosphorylation of the β_2AR is required prior to the transport from endosomes to the plasma membrane (Pippig et al., 1995; Cong et al., 2001). Since the receptor has to be palmitoylated to prevent phosphorylation in the non-activated state, palmitoylation is necessary for recycling. However, it has not been investigated whether this occurs in endosomes or at the plasma membrane.

Thus, the desensitization of β_2AR suggests a model for concerted regulation of the two post-translational modifications, with palmitoylation regulating the accessibility of phosphorylation sites involved in the desensitization of the receptor (Moffett et al., 1996). Additional GPCR have been found to be better substrates for kinases when depalmitoylated, like bradykinin B_2 receptor (Soskic et al., 1999) and rhodopsin (Karnik et al., 1993).

3.6.2 Palmitoyation for localization to lipid rafts

Many of the critical components involved in T-cell receptor-mediated signaling are localized to lipid rafts. Lipid rafts are subdomains in membranes, characterized by a resistance to extraction with cold, nonionic detergents, enriched in sphingolipids and cholesterol, and containing GPI-anchored proteins. Lipid rafts are also referred to as detergent-resistant membranes. Disruption of raft structures also disrupts early steps of T-cell receptor activation. Likewise, the mutations in the palmitoylation site of LAT (linker for activation of T cells), Fyn and Lck impair T-cell receptor-mediated signaling (Kabouridis et al., 1997; Zhang et al., 1998; van't Hof and Resh, 1999). Upon T-cell receptor activation, the T-cell receptor is phosphorylated by Fyn and Lck, which mediate a cascade of events, including phosphorylation of LAT which subsequently binds to Grb2, Gads, and phospholipase C (PLC)-γ1 via their Src homology-2 domains. Palmitoylation of Fyn, Lck and the integral membrane protein LAT, is required for their localization in lipid rafts and thereby for the T-cell receptor-mediated signaling.

Even though palmitoylation is essential for localization to lipid rafts, it is not sufficient (McCabe and Berthiaume, 2001). The GFP fused acylation sequence of the tyrosine kinase Yes

Aim of the thesis

colocalizes with cholesterol and sphingolipid-rich membranes, but not with caveolin-1, a marker for lipid rafts, while the full-length protein Yes is targeted to lipid rafts. This indicates that protein-protein interactions are required to localize proteins to lipid rafts in addition to lipid modifications.

For some viral envelope proteins, palmitoylation has also been suggested to facilitate raft-association (Nguyen and Hildreth, 2000). Rafts can provide a membrane platform on which viral structural proteins can concentrate to enhance virus assembly.

3.6.3 Palmitoylation in protein trafficking

Palmitoylation influences the trafficking of some membrane proteins. The palmitoylation deficient transferrin receptor was shown to be internalized more rapidly, but its recycling to the plasma membrane was found to be impaired, indicating that palmitoylation inhibited internalization but was required for the recycling of the receptor (Alvarez et al., 1990). In contrast, the asialoglycoprotein receptor required palmitoylation for efficient clathrin-mediated endocytosis, proper dissociation and delivery of ligand to lysosomes (Yik et al., 2002). For some membrane proteins, such as CCR5, palmitoylation is required for targeting to the plasma membrane (Blanpain et al., 2001). Palmitoylation of the CD-MPR is essential for the proper transport from endosomes to the TGN (Schweizer et al., 1996).

4 Aim of the thesis

The trafficking of the CD-MPR between the TGN and endosomes is essential for its function in the biogenesis of lysosomes. Although many sorting motifs and interacting proteins have been identified in the cytoplasmic tail of the CD-MPR, the precise mechanisms involved in the specific transport steps of the CD-MPR are not yet fully understood.

For the transport of the CD-MPR from late endosomes to the TGN, it was shown that the palmitoylation of C^{34} and the diaromatic motif $F^{18}W^{19}$ are required. The dependence of the diaromatic motif on the palmitoylation indicates that the palmitoylation induces a conformational change to better expose the motif. Our hypothesis suggests, that palmitoylation regulates the sorting signals in the cytoplasmic tail of the CD-MPR.

The aim of one project of this thesis was to answer the following questions:
- Is the palmitoylation involved in the regulation of the sorting signals of the CD-MPR?
  This question was addressed:
  a) by establishing an *in vitro* palmitoylation assay with purified full-length CD-MPR and [^{3}H]palmitoyl-CoA.
  b) by identifying the localization of the palmitoyltransferase, by testing fractions from subcellular fractionation on gradients in the *in vitro* palmitoylation assay and by *in vivo* labeling of the cells expressing wild-type or mutant CD-MPR with [^{3}H]palmitate.
  c) with attempts to clone the palmitoyltransferase.

The sorting of the CD-MPR in the TGN is mediated by GGA through binding to the DXXLL motif. AP-1 was also suggested to be involved in sorting in the TGN, by taking over the cargo from GGA by a process involving phosphorylation of GGA and CD-MPR to switch binding affinities. It was shown that the binding affinity of AP-1 to the CD-MPR was increased when the CD-MPR was phosphorylated by CK2. However, other reports claimed a phosphorylation-independent binding of CD-MPR to AP-1. Another binding analysis revealed increased affinity of phosphorylated CI-MPR peptides to GGA compared to non-phosphorylated peptides, indicating a role for phosphorylation in binding to GGA. Altogether, the reports on the requirement of phosphorylation were controversial.

Aim of the thesis

The aim of the second project of this thesis was to answer the following questions:

- Is the phosphorylation of the CD-MPR involved in its sorting in the TGN?

    This question was addressed with:

    a) *in vitro* binding of GGA1 and AP-1 to wild-type CD-MPR and the various phosphorylation mutants of the CD-MPR.

    b) *in vivo* interaction of a dominant-negative GGA1 with the phosphorylation mutants of the CD-MPR and wild-type CD-MPR.

Part I:
The Palmitoyltransferase of the Cation-Dependent Mannose 6-Phosphate Receptor cycles between the Plasma membrane and Endosomes

Jacqueline Stöckli and Jack Rohrer

1. SUMMARY 68

2. INTRODUCTION 69

3. MATERIALS AND METHODS 71

4. RESULTS 77

5. DISCUSSION 85

6. ACKNOWLEDGEMENTS 89

1. Summary

The cation-dependent mannose 6-phosphate receptor (CD-MPR) mediates the transport of lysosomal enzymes from the trans-Golgi network to the endosomes. Evasion of lysosomal degradation of the CD-MPR requires reversible palmitoylation of a cysteine residue in its cytoplasmic tail, 34 amino acids distal from its transmembrane domain. Such a distant palmitoylation of a cytoplasmic domain is rare among membrane proteins and implies drastic conformational differences between the palmitoylated and non-palmitoylated form of the receptor. Since palmitoylation is reversible and essential for correct trafficking it presents a potential regulatory mechanism for the sorting signals within the cytoplasmic domain of the CD-MPR. To characterize the palmitoyltransferase activity we established an *in vitro* palmitoylation assay using purified full length CD-MPR. We could demonstrate that palmitoylation of the CD-MPR occurs enzymatically by a membrane-bound palmitoyltransferase. In addition, analysis of the localization revealed that the palmitoyltransferase cycles between endosomes and the plasma membrane. This was identified by testing fractions from HeLa cell homogenate separated on a Percoll density gradient in the *in vitro* palmitoylation assay and further confirmed by *in vivo* labeling experiments using different treatments to block protein trafficking steps at specific sites within the cell. We identified a novel palmitoyltransferase activity in the endocytic pathway, responsible for palmitoylation of the CD-MPR. The localization of the palmitoyltransferase not only fulfills the requirement of our hypothesis to be a regulator of the intracellular trafficking of the CD-MPR but may also affect the sorting/activity of other receptors cycling through endosomes.

2. Introduction

Lysosomes are intracellular organelles containing numerous acid hydrolases and serve as a major degradative compartment in eukaryotic cells (Kornfeld, 1992; Hille-Rehfeld, 1995). Delivery of newly synthesized soluble lysosomal enzymes to lysosomes is dependent on their acquisition of mannose 6-phosphate (M6P) residues in the Golgi and the trans-Golgi network (TGN). This tag acts as a recognition signal for high-affinity binding to the M6P receptors (MPRs) in the TGN. The receptor ligand complexes then leave the TGN in clathrin-coated vesicles which fuse with acidified endosomes (Le Borgne and Hoflack, 1997). Following the pH-induced dissociation of the complexes, the lysosomal enzymes are further delivered to lysosomes whereas the receptors recycle back to the TGN to repeat this process.

The 46 kDa cation-dependent mannose 6-phosphate receptor (CD-MPR) is a type I integral membrane protein. The intracellular trafficking of the CD-MPR is directed by sorting signals located in its 67-amino acid cytoplasmic tail. Internalization is mediated by three separate internalization sequences through clathrin-coated pits: a pair of phenylalanine residues (Phe^{13}-X-X-X-X-Phe^{18}), a classical tyrosine motif (Tyr^{45}-X-X-Val^{48}) and probably a C-terminal di-leucine motif (Leu^{64}-Leu^{65}) (Johnson et al., 1990; Denzer et al., 1997). Overlapping with one of the internalization motifs is a di-aromatic motif (Phe^{18}-Trp^{19}) that binds to the MPR-tail interacting protein of 47 kDa (TIP47) and is required for the sorting of the receptor from late endosomes back to the TGN thereby preventing degradation of the CD-MPR in the lysosomes (Schweizer et al., 1997; Diaz and Pfeffer, 1998; Nair et al., 2003). In addition to the di-aromatic motif, palmitoylation of the Cys^{34} is required to avoid lysosomal degradation (Schweizer et al., 1996). Palmitoylation of Cys^{34} will anchor this portion of the cytoplasmic tail of CD-MPR to the lipid bilayer, influencing the conformation of the entire cytoplasmic tail and thereby modulating the accessibility of the sorting signals. Palmitoylation of the CD-MPR is reversible with a rapid turn-over of palmitate ($\tau_{1/2}$ ≤ 2h) (Schweizer et al., 1996). These findings led to the suggestion that palmitoylation functions to regulate the presentation of overlapping sorting signals in the cytoplasmic tail of the CD-MPR. Such a tight regulation of the signals would require that palmitoylation of the CD-MPR occur enzymatically.

Enzymatic as well as non-enzymatic palmitoylation has been described (O'Brien et al., 1987; Duncan and Gilman, 1996) although the latter might only occur *in vitro*. The two most promising candidates for such an enzymatic palmitoylation were identified in yeast, Akr1p and the Erf2p/Erf4p complex that palmitoylate the casein kinase Yck2p and Ras2p, respectively (Lobo et al., 2002; Roth et al., 2002). In addition to these two palmitoyltransferases, many palmitoyltransferase activities have been reported mostly along the exocytic pathway. Some proteins (vesicular stomatitis virus G, influenza hemagglutinin and CC chemokine receptor 5) are palmitoylated in the ER-Golgi intermediate compartment (ERGIC) or cis-Golgi (Bonatti et al., 1989; Veit and Schmidt, 1993; Blanpain et al., 2001). Further in the secretory pathway a palmitoyltransferase activity localized to the Golgi palmitoylates p21^{N-ras} (Gutierrez and Magee, 1991). For many cytosolic and membrane proteins (Fyn, edG$_\alpha$-subunits of G-protein, transferrin receptor) palmitoylation seems to occur at the plasma membrane (Adam et al., 1984; Dunphy et al., 1996; van't Hof and Resh, 1997; Fishburn et al., 1999). Furthermore, there is a palmitoyltransferase activity in mitochondria (Corvi et al., 2001) and in yeast on the vacuole (Veit et al., 2001). The various subcellular sites and the diversity of the substrates for palmitoylation suggest that numerous palmitoyltransferases with different subcellular locations and specificities may exist.

Three thioesterases which remove the palmitate residue form proteins were identified so far in mammals (Camp and Hofmann, 1993; Soyombo and Hofmann, 1997a; Duncan and Gilman, 1998). Two of these thioesterases are soluble lysosomal enzymes, palmitoyl-protein thioesterase 1 and 2 (PPT1, PPT2), required for degradation of palmitoylated substrates (Verkruyse and Hofmann, 1996; Soyombo and Hofmann, 1997a). The third thioesterase, acyl protein thioesterase 1 (APT1) is a cytoplasmic protein and was shown to depalmitoylate substrates, such as G protein α subunits, p21^{ras} and endothelial nitric-oxide synthase (Duncan and Gilman, 1998; Yeh et al., 1999). Whether the CD-MPR gets enzymatically palmitoylated and/or depalmitoylated has not been investigated so far.

In order to confirm or overthrow our hypothesis that reversible palmitoylation of the CD-MPR is regulating its sorting a detailed characterization of its palmitoylation is required.

3. Materials and Methods

Materials

Enzymes used in molecular cloning were obtained from Roche Diagnostics (Mannheim, Germany), New England Biolabs (Beverly, MA, USA), or Promega (Madison, WI, USA); general chemicals from Fluka (Buchs, Switzerland); protease inhibitors, coenzyme A and wortmannin from Sigma (St. Louis, MO, USA); Dulbecco's Modified Eagle Medium (DMEM), fetal calf serum (FCS), G418 and Lipofectamine Plus were from Invitrogen (Carlsbad, CA, USA); cell culture dishes from Falcon (Franklin Lakes, NJ, USA); nitrocellulose from Schleicher & Schuell (Dassel, Germany); enhanced chemiluminescence Western blotting reagents from PerkinElmer Life Sciences (Boston, MA, USA); protein A-Sepharose beads from Repligen Corp. (Cambridge, MA, USA); Percoll, activated CH Sepharose 4B and low molecular weight protein markers from Amersham Pharmacia Biotech (Piscataway, NJ, USA); Centricon Plus-20 from Millipore Corporation (Bedford, MA, USA); [^{3}H]palmitate from ARC (St. Louis, MO, USA) or from ANAWA Trading SA (Zürich, Switzerland); Acyl-CoA synthetase from Fluka (Buchs, Switzerland). Oligonucleotides were synthesized either by the DNA synthesis facility of the Friedrich Miescher Institute (Basel, Switzerland) or Microsynth GmBH (Balgach, Switzerland).

Antibodies

Rabbit anti-mouse IgG was purchased from Zymed Laboratories, Inc. (San Francisco, CA, USA). Horseradish peroxidase conjugated antibodies against mouse and rabbit were from Amersham Pharmacia Biotech (Piscataway, NJ, USA). Alexa 488 conjugated goat anti-mouse antibody was from Molecular Probes (Eugene, OR, USA). The monoclonal antibody 22D4 specific for the bovine CD-MPR was generously provided by D. Messner (Messner, 1993). This monoclonal antibody is specific for the bovine CD-MPR and does not cross-react with the endogenous mouse CD-MPR. The monoclonal antibodies for NaKATPase (N1/123/33) and p63 (G1/296/22) were a gift from H.P. Hauri (Marxer et al., 1989; Schweizer et al., 1993a). The monoclonal antibody for Rab5 was generously provided by J. Gruenberg (Gorvel et al., 1991) and the polyclonal antibody for Rab7 (no 14-1) was kindly supplied by P. Chavrier (Chavrier et al., 1990). The monoclonal antibody for β1-4-galactosyltransferase-1 was a gift from E. Berger (Berger et al., 1986).

Recombinant DNA

All basic DNA procedures were as described (Sambrook et al., 1998). The PCR procedure of Ho and colleagues (Ho et al., 1989) was used to generate the MPR-FFWYLL-A construct with pSFFV-MPR (Rohrer et al., 1995) serving as a template together with MPR-BglII.down (5'-CCGAGATCTCCCACTTAAGCGTGG-3') and pSFFVneo.up2 (5'-CTGCCATTCATCCGCTTATTATC-3') as the down- and upstream primers, respectively. Appropriate partial complementary pairs of oligonucleotides in which the desired amino acid replacement had been incorporated were chosen as internal primers. The final PCR product was subcloned into pSFFVneo as described (Rohrer et al., 1995) and confirmed by sequencing.

Cell Culture and Transfection

A mannose-6-P/insulin-like growth factor-II receptor-deficient mouse L cell line designated D9 (LRec–) was maintained in DMEM containing 10% FCS. The cells were transfected with XbaI-linearized DNA with Lipofectamine Plus according to the manufacturer's directions. Selection for resistance to neomycin (G418) was carried out using 500 µg/ml G418 as the final concentration. Resistant colonies were picked individually and screened for the expression of bovine CD-MPR by immunoblotting. Clones expressing similar amounts of receptor compared to ML4 cells, the reference cell line expressing wt bovine CD-MPR (Johnson et al., 1990), were expanded for further study and maintained in selective medium.

Internalization Assay

Cells grown in 6-well plates were chilled on ice, rinsed four times with ice-cold PBS and then incubated with 1.5 ml of 3 mg/ml sulfo-NHS-SS-biotin in PBS for 15 min to biotinylate surface proteins. Biotinylation was stopped by washing once with 50 mM glycine in PBS and twice with PBS. Some of the cells were then incubated at 37 °C with prewarmed growth medium containing 10% fetal calf serum and 20 mM Hepes, pH 7.4, for different periods of time (1, 5, and 15 min). The cells were returned to 4 °C to stop internalization, washed once with ice-cold PBS and then incubated on ice twice for 20 min in a freshly prepared glutathione solution (50 mM L-glutathione reduced, 75 mM sodium chloride, 1 mM EDTA, pH 8, 0.075 N NaOH, 1% BSA) to remove the biotin from proteins that were present on the cell surface. In addition, two samples that were not incubated at 37 °C were treated either with (0% control) or without (100% control) the glutathione solution on ice. After reduction, the cells were washed with PBS and the excess glutathione was quenched with a 5 min incubation on ice in PBS containing 5 mg/ml iodoacetamide. The cells were then washed twice with PBS, lysed in 1 ml buffer 2 (100 mM sodium phosphate (pH 8.0), containing 1% Triton X-100 and a 1:500 dilution of a protease inhibitor cocktail (5 mg/ml benzamidine, and 1 mg/ml each of pepstatin A, leupeptin, antipain, and chymostatin in 40% dimethyl sulfoxide-60% ethanol) (PIC) and phenylmethylsulfonyl fluoride (PMSF) (40 µg/ml) and passed five times through a 25-gauge needle connected to a 1-ml syringe. After solubilizing for 30 min on ice, the cell lysates were centrifuged for 30 min at 40'000 rpm in a Ti 50 rotor (Beckman Instruments Inc.). The resulting supernatants were subjected to immunoprecipitation and Western blotting as described below. The biotinylated fraction of the proteins was then detected by enhanced chemiluminescence using streptavidin-horseradish peroxidase.

Steady State Surface Distribution of CD-MPR

Confluent cells in 12-well plates were washed with PBS and incubated for 15 min on ice with either 0.5 ml 10 mg/ml BSA in PBS (cell surface) or with 0.5 ml PBS containing 10 mg/ml BSA and 0.1% saponin (total). The cells were incubated with 5×10^5 cpm of ^{125}I-labeled 22D4 mAb against the CD-MPR in either 300 µl 10 mg/ml BSA in PBS (cell surface) or 300 µl PBS containing 10 mg/ml BSA and 0.1% saponin (total) on ice. After 2h, the cells were washed five times with 1 ml 10 mg/ml BSA in PBS or 10 mg/ml BSA/0.1% saponin in PBS and solubilized in 0.5 ml of 0.1 M NaOH. Cell-associated radioactivity was determined with a γ-counter. The iodination of the antibody was performed by ANAWA Trading SA (Zürich, Switzerland) according to their standard protocol.

Metabolic Labeling with [³H]Palmitate

Cells were grown in 6-well plates. For the treatment at 19°C the cells were washed twice with PBS and pretreated at 19°C or 37°C for 30 min in 1 ml DMEM containing 20 mM Hepes (pH 7.4) and 5% low-lipid calf serum. For the treatment with wortmannin the cells were washed twice with PBS and preincubated with DMEM containing 20 mM Hepes (pH 7.4), 5% low-lipid calf serum and 1 µM wortmannin or 0.1% DMSO for control cells for 45 min at 37°C. After preincubation all cells were incubated with 150 µCi [³H]palmitate in 1 ml preincubation medium for 90 min at 37°C. For the wortmannin treatment 1 µM wortmannin was added again to the labeling media after 45 min. After labeling, cells were chilled on ice, washed once with ice-cold PBS, scraped in 1 ml ice-cold PBS and centrifuged for 5' at 260 x *g* at 4°C. The pellets were lysed in 1 ml buffer-2 (100 mM sodium phosphate (pH 8.0), containing 1% Triton X-100 and a 1:500 dilution of a protease inhibitor cocktail (5 mg/ml benzamidine, and 1 mg/ml each of pepstatin A, leupeptin, antipain, and chymostatin in 40% dimethyl sulfoxide-60% ethanol) (PIC) and phenylmethylsulfonyl fluoride (PMSF) (40 µg/ml) and passed five times through a 25-gauge needle connected to a 1-ml syringe. After solubilizing for 30 min on ice, the cell lysates were centrifuged for 30 min at 40'000 rpm in a Ti50 rotor (Beckman Instruments Inc.). The resulting supernatants were subjected to immunoprecipitation and subsequently analyzed by SDS-PAGE and fluorography as described below.

Synthesis of [^{3}H]Palmitoyl-CoA

[^{3}H]palmitoyl-coenzyme A ([^{3}H]palmitoyl-CoA) was prepared from 750 µCi [^{3}H]palmitate (60Ci/mmol, 10mCi/ml) by incubation with 0.05 U acyl-CoA synthetase in 1 ml of 0.05% Triton X-100, 0.5 mM coenzyme A, 1 mM ATP, 1 mM MgCl$_2$, 40 mM KH$_2$PO$_4$, pH 7.5 for 1 hour at 37°C. The sample was dried in a Speed-Vac centrifugal evaporator (Savant Instr. Inc., Hicksville, NY, USA) and resuspended in 75 µl 50 mM Tris, pH 8.0. The purity was determined by thin-layer chromatography on Merck Silica Gel 60 plate using propanol-water-5% ammonia (70:10:20) as the developing solvent, resulting in an R_f of 0.43 for palmitoyl-CoA and 0.61 for palmitate. The analysis showed ≥90% radiochemical purity.

CD-MPR purification

Mouse L cells stably expressing bovine CD-MPR wt were grown in suspension. 1.5x10^9 cells were centrifuged for 5 min at 260 x g at 4°C, washed with PBS, lysed in 8 ml buffer-2 containing 1:500 dilution of a protease inhibitor cocktail and sonicated. After solubilizing for 30 min on ice, the cell lysates were centrifuged for 30 min at 40'000 rpm at 4°C in a Ti 50 rotor (Beckman Instruments Inc.). The supernatant was loaded onto a 22D4 antibody column. This column was prepared by coupling 5 mg purified 22D4 monoclonal antibody to 5 ml activated CH-Sepharose 4B according to the manufacturer's protocol. The column was washed with 150 ml buffer-2 containing protease inhibitors at a flow rate of 1 ml/min. Bound CD-MPR was eluted with elution buffer (0.1 M glycine pH 3.0, 0.05% Triton X-100) in 1 ml fractions containing 100 µl 1 M Tris, pH 8.2 to neutralize. 10 µl of each fraction were subjected to SDS-PAGE and immunoblotting with 22D4 antibody. The fractions containing CD-MPR were pooled and concentrated with Centricon Plus-20 to obtain a concentration of 1 mg/ml. The Triton X-100 content was measured in a spectrophotometer at 277 nm wavelength and the protein concentration was determined using a Bio-Rad protein assay. The sample was aliquoted and frozen at -20°C.

Percoll Density Gradient

HeLa cells were grown on 15 cm culture dishes. Cells from twelve dishes were washed with PBS and scraped in 5 ml homogenization buffer (10 mM Tris, pH 7.4, 0.25 M sucrose). The cells were centrifuged at 260 x g for 5 min at 4°C, resuspended in 5 ml homogenization buffer containing protease inhibitors and homogenized in a ball-bearing homogenizer (HMG, Heidelberg, Germany) with 12 strokes. The homogenate was centrifuged at 700 x g for 10 min at 4°C and 5 mg of the resulting post-nuclear supernatant (PNS) were diluted in 2.3 ml homogenization buffer and mixed with 9.2 ml 15% Percoll. The resulting 12% Percoll sample was loaded on top of a 0.5 ml 2.5 M sucrose pillow into a thin-walled open-top centrifugation tube and centrifuged at 20'000 rpm (28'000 x g) in a Ti70.1 rotor for 45 min at 4°C. 1 ml fractions were collected from the bottom of the gradient and the membranes of each fraction were pelleted by centrifugation at 80'000 rpm in a TLA 120.2 rotor for 30 min at 4°C and resuspended in equal volumes. The fractions were assayed for β-hexosaminidase activity (lysosomal enzyme) and subjected to SDS-PAGE and immunoblotting with antibodies against NaK-ATPase, galactosyltransferase, p63, Rab5 and Rab7 to determine the expression levels of these marker enzymes.

In vitro Palmitoyltransferase Assay

10 μg purified CD-MPR and 200 μCi [^{3}H]palmitoyl-CoA were incubated with 150 μg protein from HeLa cell PNS or membrane fraction in a total volume of 500 μl assay buffer (45 mM Tris, pH 8.0, 40.5 mM glycine, 2 mM ATP, 130 mM KCl, 10 mM NaCl, 1 mM DTT, 0.02% Triton X-100). The sample was incubated at 37°C for 30 min. 500 μl 2x buffer-2 containing PIC and PMSF was added to the sample and solubilized on ice for 30 min followed by a centrifugation at 100'000 x g in a Ti50 rotor for 30 min at 4°C. The resulting supernatants were subjected to immunoprecipitation and subsequently analyzed by SDS-PAGE and fluorography as described below.

Immunoprecipitation, SDS-PAGE, Fluorography and Immunoblotting

For immunoprecipitation with anti-CD-MPR mAb 22D4, 30 μl of protein A-Sepharose was washed once with 1 ml buffer-1 (100 mM sodium phosphate (pH 8.0), 0.2% BSA) and then incubated with 1.5 μl rabbit anti-mouse antibody (1 mg/ml) in 500 μl buffer-1 for 2h at 4°C. After two washes with buffer-1, the beads were incubated with 10 μl of mAb 22D4 in 500 μl buffer-1 for 2h at 4°C. The beads were then washed with 1 ml buffer-1 and with 1 ml buffer-2. The washed beads and the radiolabeled cell lysates were combined and incubated overnight at 4°C with constant mixing. The protein A-Sepharose beads were pelleted, washed three times with buffer-2, and then once with 100 mM sodium phosphate (pH 8.0), followed by a final wash step with 10 mM sodium phosphate (pH 8.0). The immunocomplexes were released from the beads by boiling for 3 min in non-reducing SDS-PAGE sample buffer (94 mM Tris-HCl (pH 6.8), 3% SDS, 15% glycerol, 0.001% bromophenol blue). The proteins were separated on a 10% SDS-polyacrylamide minigel by using the Laemmli system (Laemmli, 1970). For fluorography the gel was stained with 0.25 g Coomassie brilliant blue R-250 in 100 ml destaining solution (25% methanol, 7% acetic acid in

H_2O) for 20 min, destained for 1h in destaining solution, incubated in H_2O for 5 min, treated with 1 M sodium salicylate for 20 min, dried and exposed to XOmat AR film (Kodak, Eastman Kodak Company, Rochester, NY, USA) for 3 to 10 days. For Western blotting the gel was transferred onto a nitrocellulose membrane according to the method of Towbin (Towbin et al., 1979). The membrane was blocked with 3% nonfat dry milk powder (Sano Lait, Coop, Switzerland) in PBS. The blot was subsequently incubated with mAb 22D4 (diluted 1:500 in PBS-3% powdered milk) followed by a horseradish peroxidase conjugated anti-mouse secondary antibody (diluted 1:2000 in PBS-3% powdered milk). Immunoreactive proteins were visualized using the enhanced chemiluminescence detection system according to the manufacturer's directions. The autoradiographs were quantitated using a personal densitometer (Amersham Pharmacia Biotech).

Confocal Immunofluorescence Microscopy

Cells were grown on coverslips, washed with PBS and fixed in 3% paraformaldehyde pH 8.3 for 20 min followed by four washes with 20 mM glycine in PBS. The cells were permeablized in saponin buffer (0.1% saponin, 20 mM glycine in PBS) for 20 min. All following steps were performed in saponin buffer. Cells were incubated with 22D4 antibody (1:500) for 30 min and washed four times followed by incubation with goat anti-mouse Alexa 488 antibody. The coverslips were washed four times and mounted on glass slides with ProLong Antifade (Molecular Probes, Eugene, OR, USA) for viewing with a Leica SP2 AOBS UV confocal laser-scanning microscope. Serial sections in the z axis through the entire cells were taken, and the resulting stacks of images were analyzed with the use of the Imaris program (Bitplane AG, Zürich, Switzerland).

Assays and Miscellaneous Methods

β-Hexosaminidase activity was determined by dilution of the samples in 300 µl 1.67 mM p-nitrophenyl N-acetyl-β-glucosaminide, 50 mM citrate, pH 4.0, 0.1% Triton X-100. The samples were incubated for 30-60 min at 37°C. The reaction was stopped with 1 ml 0.2 M sodium carbonate, and the absorbance read at 400 nm. Protein concentration was determined with the Bio-Rad (Hercules, CA, USA) protein assay kit using protein standard I according to the manufacturer's protocol.

4. Results

CD-MPR is enzymatically palmitoylated

CD-MPR is reversibly palmitoylated at Cys^{30} and Cys^{34}, wherein palmitoylation of the Cys^{34} is essential for the trafficking of the CD-MPR from endosomes to the TGN (Schweizer et al., 1996). To explore the nature of this reversible palmitoylation of the CD-MPR, an *in vitro* palmitoylation assay was developed. 10 µg of CD-MPR, purified as a full length membrane protein, was incubated with 200 µCi [^{3}H]palmitoyl-CoA for 30 min at 37°C with HeLa cell homogenate. To maintain the

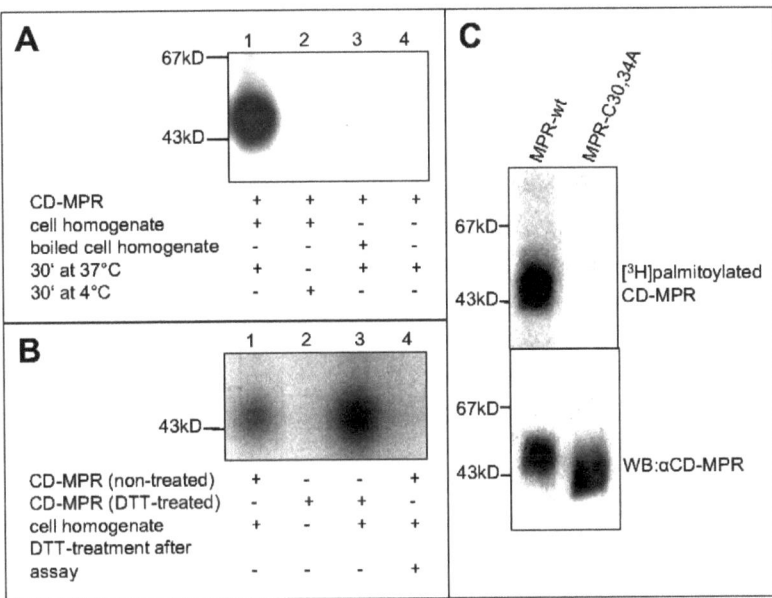

Figure 8: CD-MPR palmitoylation occurs enzymatically. *A*, 10 µg purified CD-MPR was incubated with [3H]palmitoyl-CoA and 150 µg HeLa cell homogenate for 30 min at 37°C (lane 1) or at 4°C (lane 2). In lane 3, 150 µg HeLa cell homogenate was boiled for 15 min prior to incubation with CD-MPR and [3H]palmitoyl-CoA for 30 min at 37°C. Lane 4 shows the sample of an assay without HeLa cell homogenate. CD-MPR was immunoprecipitated and subjected to SDS-PAGE (10% gel). [3H]palmitate incorporation into CD-MPR was visualized by autoradiography. *B*, 10 µg purified CD-MPR was incubated with [3H]palmitoyl-CoA and 150 µg HeLa cell homogenate for 30 min at 37°C (lane 1). 10 µg purified CD-MPR was incubated with 50 mM DTT for 2 h at 50°C prior to incubation with [3H]palmitoyl-CoA and either without (lane 2) or with 150 µg HeLa cell homogenate (lane 3) for 30 min at 37°C. CD-MPR was immunoprecipitated. The sample in lane 4 corresponds to the sample in lane 1 that was subsequently incubated with 50 mM DTT for 2h at 50°C. The samples were separated on SDS-PAGE (10% gel). [3H]palmitate incorporation into CD-MPR was visualized by autoradiography. *C,* post-nuclear supernatant (PNS) of mouse L cells stably transfected with CD-MPR wild-type (wt) or CD-MPR C30,34A were incubated with [3H]palmitoyl-CoA for 30 min at 37°C. CD-MPR wild-type and mutant were subsequently immunoprecipitated and subjected to SDS-PAGE (10% gel). [3H]palmitate incorporation into CD-MPR was visualized by autoradiography (upper panel). Expression levels of the CD-MPR wild-type and C30,34A in the PNS were determined by Western blotting (WB) with the anti-CD-MPR monoclonal antibody 22D4 (lower panel).

CD-MPR in solubilized form, 0.02% Triton X-100 was used. In a first step the assay was validated by demonstrating that palmitoylation of the CD-MPR occurs enzymatically. Incubation of purified CD-MPR with [^{3}H]palmitoyl-CoA alone did not result in palmitoylated CD-MPR in contrast to the incubation of the receptor with [^{3}H]palmitoyl-CoA and HeLa cell extract (Figure 8A, lane 1 versus lane 4). Next, a potential palmitoyltransferase was either denatured by boiling the HeLa cell homogenate for 15 min prior to the incubation at 37°C, or alternatively the entire reaction was carried out at 4°C (Figure 8A, lanes 2, 3). In both cases no palmitoylated CD-MPR was detected. Furthermore, experiments with varying time and temperature of the incubation revealed that the assay is dependent on those parameters (data not shown). Altogether, these findings demonstrate that palmitoylation of the CD-MPR in this assay requires an active enzyme. Moreover, no autocatalytic palmitoylation was obtained when purified CD-MPR was treated with 50 mM dithiothreitol (DTT) for 2h at 50°C to hydrolyze already attached palmitate moieties prior to the incubation with [^{3}H]palmitoyl-CoA (Figure 8B, lane 2). However, DTT-treated CD-MPR was readily palmitoylated when HeLa cell homogenate was added to the assay (Figure 8B, lane 3). This showed that the enzymatic requirement was specific for palmitoylation and not the hydrolysis of the palmitate. The palmitic acid was attached via a thioester linkage which was demonstrated by treating the [^{3}H]palmitoylated CD-MPR following the in vitro palmitoylation assay with 50 mM DTT for 2h at 50°C. The [^{3}H]palmitate was released by DTT that hydrolyses thioesters (Figure 8B, lanes 1 and 4), confirming previous results *in vivo* (Schweizer et al., 1996). In order to show, that the palmitoyltransferase in the *in vitro* assay specifically palmitoylated the cytoplasmic cysteines and not the lumenal ones, we used post-nuclear supernatant (PNS) from mouse L cells stably transfected with wt CD-MPR and with a mutant CD-MPR, with both cytoplasmic cysteines replaced by alanines (C30,34A). In this experiment the PNS comprised the palmitoyltransferase activity, as well as the appropriate substrate, which was immunoprecipitated after incubation with [^{3}H]palmitoyl-CoA. We could show that the lumenal cysteines of the CD-MPR in the C30,34A were not substrate for the palmitoyltransferase, since this mutant did not show incorporation of [^{3}H]palmitate (Figure 8C).

Palmitoyltransferase is a membrane protein

The localization of the palmitoyltransferase was investigated by fractionating the cell homogenate. The 700 x *g* supernatant (post nuclear supernatant (PNS)) was subjected to a 100'000 x *g* spin. The pellet containing the membranes and the supernatant containing the cytosol were both tested for palmitoyltransferase activity in the *in vitro* assay. The vast majority of the activity was found in the pellet, thus in the membrane fraction (Figure 9). To differentiate between peripheral and integral membrane proteins, the pellet of the 100'000 x *g* spin was incubated with 0.5 M NaCl for 30 min at 4°C and then centrifuged once more at 100'000 x g for 30 min at 4°C. The resulting supernatant contained peripheral membrane proteins that were detached from the membrane, whereas the pellet consisted of integral membrane proteins. The CD-MPR palmitoyltransferase activity determined in the *in vitro* assay was recovered in the pellet, although reduced, which might reflect the exposure of the palmitoyltransferase to high salt concentration (Figure 9). Furthermore, a carbonate wash was applied to the membrane fraction. This harsher treatment more accurately separates peripheral from integral membrane proteins. The sample was incubated with 0.1 M sodium carbonate pH 11 for 30 min at 4°C, neutralized with HCl and subsequently centrifuged at 100'000 x g for 30 min at 4°C. When tested in the *in vitro* assay, no palmitoylated CD-MPR was detected in the pellet or in the supernatant (data not shown), indicating that the exposure to high pH irreversibly destroyed the palmitoyltransferase activity.

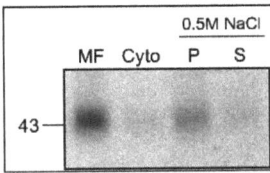

Figure 9: Palmitoyltransferase is membrane-bound. The post-nuclear supernatant of HeLa cells was centrifuged at 100'000 x g for 30 min at 4°C. The pellet contained the membrane fraction (MF) and the supernatant consisted of the cytosol (Cyto). For the salt wash the membrane fraction was subsequently incubated with 0.5 M NaCl for 30 min at 4°C and centrifuged again. The resulting pellet (P) and the supernatant (S) as well as the membrane fraction (MF) and the cytosol (Cyto) were assayed for palmitoyltransferase activity *in vitro*. The samples were incubated with [^{3}H]palmitoyl-CoA and purified CD-MPR at 37°C for 30 min. CD-MPR was immunoprecipitated und subjected to SDS-PAGE. [^{3}H]palmitate incorporation into CD-MPR was visualized by autoradiography.

Cell fractionation to localize the palmitoyltransferase activity

To determine the localization of the palmitoyltransferase, HeLa cell homogenate was separated on a self forming Percoll density gradient and 12 individual fractions were harvested from the bottom. The fractions were then centrifuged at 100'000 x g to pellet the membranes and subsequently resuspended in equal volumes. With this step we isolated the membranes and organelles free fromcytosol, and therefore they didn't contain the thioesterase APT1 which might act on the CD-MPR and thus influence the experimental data. The same percent of the fractions of the gradient was tested for expression of marker enzymes of the different organelles. A typical result is shown in Figure 10. The gradient clearly separated peaks of β-hexosaminidase (lysosomes), p63 (rough endoplasmic reticulum), galactosyltransferase (Golgi) and NaK-ATPase (plasma membrane) (Figure 10A and C). The distribution of marker proteins for the early (Rab5) and late endosomes (Rab7) was not as clearly separated as the other marker proteins. Rab5 had two major peaks, one very dense peak close to the lysosomal β-hexosaminidase (fraction 2), reflecting a heavy, probably coated early endosomal structure and a broad peak partially coinciding with the marker of the plasma membrane, representing a light early endosomal structure. The majority of Rab7 was found on a dense fraction (fraction 3) trailing off into lighter fractions (fractions 4-8), reflecting the heterogeneous nature of late endosomes. For each fraction 150 µg of protein was tested in the *in vitro* assay, to determine the specific palmitoyltransferase activity (Figure 10B). The specific activity of the palmitoyltransferase was quantified, subsequently the total activity per fraction was calculated and plotted in a graph together with the marker proteins for comparison (Figure 10C). The CD-MPR palmitoyltransferase activity showed a distribution with several small peaks that coincided with the distribution of the early endosomes, possibly reflecting an association of the palmitoyltransferase with this organelle. Alternatively, the palmitoyltransferase could be

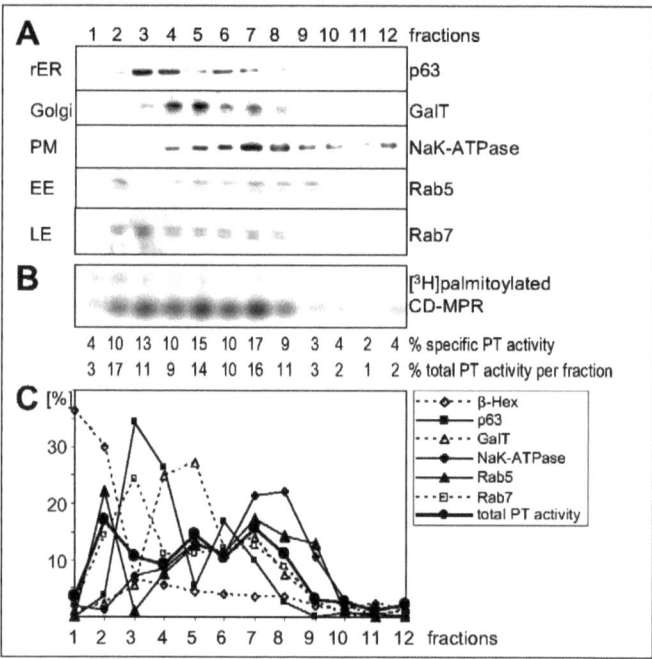

Figure 10: HeLa cell fractionation assayed for palmitoyltransferase (PT) activity *in vitro*. *A*, The postnuclear supernatant of HeLa cells was separated on a 15% percoll density gradient. Fractions were collected from bottom (fraction 1) to top (fraction 12) of the gradient. Membranes of fractions were centrifuged and pellets were resuspended in equal volumes. Equal volumes of fractions were tested for expression of marker enzymes of the different organelles by immunoblotting with the respective antibodies: galactosyltransferase (GalT) for Golgi, NaK-ATPase for plasma membrane (PM), p63 for rough endoplasmic reticulum (rER), Rab7 for late endosomes (LE) and Rab5 for early endosomes (EE). *B*, 150 µg of each fraction was assayed for palmitoyltransferase activity *in vitro*. The samples were incubated with [^{3}H]palmitoyl-CoA and purified CD-MPR at 37°C for 30 min. CD-MPR was immunoprecipitated and subjected to SDS-PAGE. [^{3}H]palmitate incorporation into CD-MPR was visualized by autoradiography. The percentage of specific activity in each fraction was quantitated by densitometric scanning. The total activity per fraction was calculated using the protein concentration of each fraction. *C*, Graph combining percentages of marker enzyme expression per fraction and total palmitoyltransferase activity per fraction. Immunoblot shown in *A* was quantitated by densitometric scanning. Levels of β-hexosaminidase, a lysosomal marker enzyme, were measured with an activity assay with equal volumes of fractions.

localized to several organelles including the Golgi, the plasma membrane and heavy early endosomal structures (Figure 10C).

In vivo labeling with [³H]palmitate to localize palmitoyltransferase activity

Further investigation to localize the palmitoyltransferase was performed with *in vivo* labeling experiments by adding [³H]palmitate to the cells which was incorporated by the CD-MPR. These experiments were carried out with two CD-MPR constructs, the wt and a mutant construct (FFWYLL-A, Figure 11A), stably transfected into mouse L cells. In the mutant CD-MPR the important amino acids of all three internalization signals (Denzer et al., 1997) were mutated to alanines. Therefore the mutant CD-MPR was not internalized, whereas the wt CD-MPR was rapidly internalized Figure 11B). This led to an accumulation of the mutant CD-MPR at the plasma

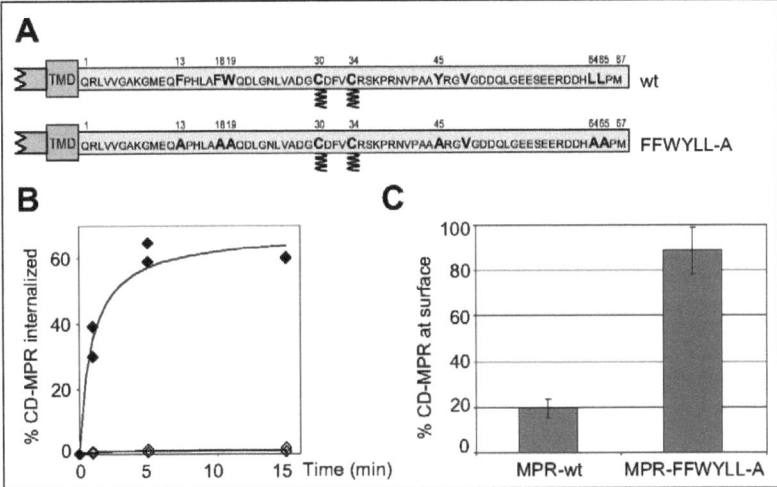

Figure 11: **Increased surface levels and decreased internalization rate of MPR-FFWYLL-A.** *A*, Schematic illustration of the cytoplasmic tails of the CD-MPR constructs. The amino acids are shown in single letter code. The internalization signals which are mutated to alanines in the mutant construct and the palmitoylated cysteines are indicated by bold letters. *B*, Cell surface proteins of mouse L cells stably expressing MPR wt (♦) and MPR-FFWYLL-A (◇) were derivatized at 4°C using sulfo-NHS-SS-biotin. The cells were then incubated at 37 °C for the indicated time and subsequently chilled on ice. The biotin groups remaining at the cell surface were removed by incubation in a reducing glutathione solution. The cells were lysed, and the wild type and mutant form of MPR were immunoprecipitated. Immunoprecipitates were resolved by SDS-PAGE and subjected to immunoblotting using a streptavidin-horseradish peroxidase conjugate. The immunoblots of two experiments were quantitated for each construct, and the values were expressed as their percentage of the sample that was kept at 4 °C and not treated with glutathione. The values were shown from two separate experiments. *C*, Mouse L cells stably expressing MPR wt and MPR-FFWYLL-A were incubated with iodinated antibodies against CD-MPR for 2 hours on ice either without saponin for the surface levels or with 0.1% saponin to determine the total CD-MPR levels. The cells were lysed and the cell-associated radioactivity was determined with a γ-counter. The bars represent the percentage of wt and mutant CD-MPR that were present at the cell surface at steady state. The values are expressed as mean ± S.E.M. from four separate experiments.

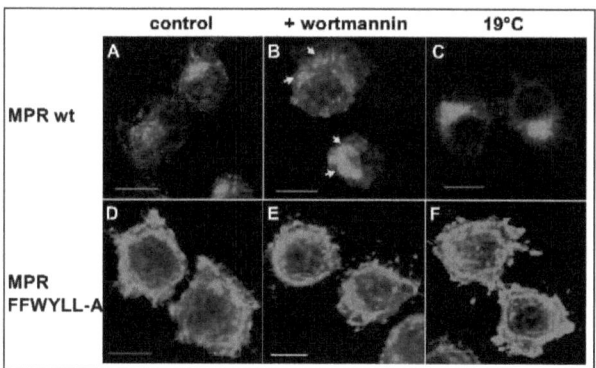

Figure 12: Effect of wortmannin and 19°C temperature block on the localization of CD-MPR wt and FFWYLL-A. Mouse L cells stably expressing MPR wt (*A-C*) and MPR-FFWYLL-A (*D-F*) were fixed, permeabilized and incubated with the monoclonal antibody 22D4 against CD-MPR followed by goat anti-mouse Alexa 488. Prior to fixation *B* and *E* were treated with 1 μM wortmannin for 90 min, whereas *C* and *F* were incubated at 19°C for 90 min. Scale bars are 10 μm. Arrows in *B* indicate enlarged endosomes.

membrane of 88%, while wt CD-MPR was expressed at the plasma membrane only to 19% (Figure 11C). The cells were subjected to two treatments to block certain transport steps within the cell, in order to accumulate the proteins in a specific organelle. The levels of palmitoylated CD-MPR in control cells and treated cells were compared to determine whether the palmitoyltransferase and CD-MPR accumulated in the same organelle. One block was evoked by wortmannin, a fungal

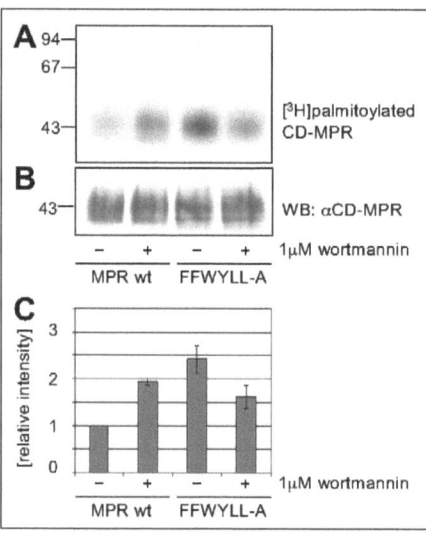

Figure 13: CD-MPR palmitoylation upon wortmannin treatment *in vivo*. *A*, Mouse L cells stably expressing MPR wt and MPR-FFWYLL-A were pretreated with 1 μM wortmannin or 0.1% DMSO for 45 min and then labeled with [^{3}H]palmitate for 90 min in the presence of 1 μM wortmannin (repeatedly added after every 45 min) or 0.1% DMSO. The cells were then chilled on ice and lysed. CD-MPR was immunoprecipitated and subjected to SDS-PAGE. [^{3}H]palmitate incorporation into CD-MPR was visualized by autoradiography. *B*, Level of CD-MPR expression. 10% of the immunoprecipitated sample was subjected to SDS-PAGE and Western blotting (WB) with 22D4 mAb against CD-MPR. *C*, Quantitation of [^{3}H]palmitate incorporation into CD-MPR wt and FFWYLL-A. The fluorography and the Western blot shown in *A* and *B*, respectively, and those from additional experiments were quantitated by densitometric scanning. In each experiment the values obtained for the [^{3}H]palmitate incorporation were corrected for the CD-MPR expression levels. The value obtained with the DMSO treated CD-MPR wt was set to 1. The values are expressed as mean ± S.E.M. from three separate experiments.

metabolite which inhibits the phosphatidylinositol 3-OH kinase (PI3K). It has been shown that the addition of wortmannin to cells at micromolar concentration for one hour causes an accumulation of the cation-independent mannose 6-phosphate receptor (CI-MPR) and furin in enlarged endosomes (Brown et al., 1995; Kundra and Kornfeld, 1998; Mallet and Maxfield, 1999). In order to accumulate proteins in endosomes, cells were pretreated with 1 µM wortmannin or as a control with 0.1% dimethyl sulfoxide (DMSO) for 45 min and labeled with [^{3}H]palmitate for 90 min at 37°C in the presence of 0.1% DMSO or wortmannin which was repeatedly added every 45 min due to the short half-life of wortmannin. Upon treatment with wortmannin, wt CD-MPR accumulated in enlarged endosomes (Figure 12A, B arrows), while the mutant CD-MPR at the plasma membrane had no altered distribution (Figure 12D versus E). Wt CD-MPR had an increased level of palmitoylation upon addition of wortmannin compared to control cells, whereas the mutant CD-MPR was less palmitoylated than in control cells (Figure 13). This result demonstrated that the palmitoyltransferase is not only present at the plasma membrane (high level of palmitoylation of the mutant CD-MPR) but also in the endosomes. Upon treatment with wortmannin, the palmitoyltransferase accumulated in the endosomes, where also wt CD-MPR accumulated and hence the level of palmitoylation of wt CD-MPR was highly elevated. Furthermore, palmitoylation of the mutant CD-MPR decreased upon treatment with wortmannin, indicating that the palmitoyltransferase was depleted from the plasma membrane. Altogether, these data show that the

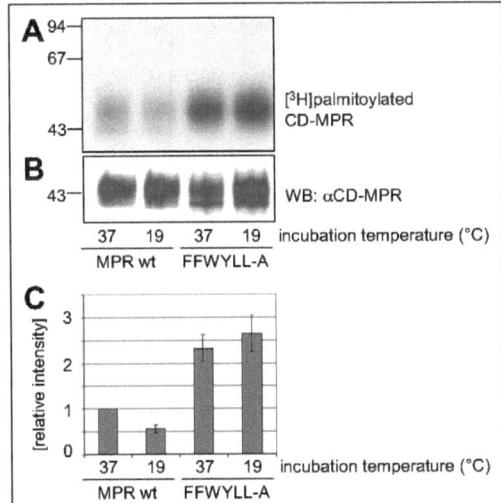

Figure 14: CD-MPR palmitoylation upon 19°C temperature block *in vivo*. *A*, Mouse L cells stably expressing MPR wt and MPR-FFWYLL-A were preincubated at 19°C or 37°C for 30 min and then labeled with [^{3}H]palmitate for 90 min at 19°C or 37°C, respectively. The cells were then chilled on ice and lysed. CD-MPR was immunoprecipitated and subjected to SDS-PAGE. [^{3}H]palmitate incorporation into CD-MPR was visualized by autoradiography. *B*, Level of CD-MPR expression. 10% of the immunoprecipitated sample was subjected to SDS-PAGE and Western blotting (WB) with 22D4 mAb against CD-MPR. *C*, Quantitation of [^{3}H]palmitate incorporation into CD-MPR wt and FFWYLL-A. The fluorography and the Western blot shown in *A* and *B*, respectively, and those from additional experiments were quantitated by densitometric scanning. In each experiment the values obtained for the [^{3}H]palmitate incorporation were corrected for the CD-MPR expression levels. The value obtained with the CD-MPR wt incubated at 37°C was set to 1. The values are expressed as mean ± S.E.M. from three separate experiments.

palmitoyltransferase cycles between the plasma membrane and endosomes and that the addition of wortmannin trapped the enzyme in endosomes. To further investigate the internal trafficking pathway of the palmitoyltransferase, a different block of intracellular transport was applied. Incubation of cells at 19°C blocks the exit out of the TGN, therefore accumulating proteins that cycle through the TGN (Matlin and Simons, 1983; Griffiths et al., 1985). Cells were preincubated either at 19°C or 37°C for 30 min and subsequently labeled with [^{3}H]palmitate for 90 min at 19°C or 37°C. As expected, the temperature block accumulated wt CD-MPR in the TGN, whereas the localization of the mutant CD-MPR at the plasma membrane was not affected (Figure 12A,C,D,F). The level of palmitoylation of the mutant CD-MPR was not changed by the temperature block, whereas the extent of palmitoylation of the wt CD-MPR was strongly reduced at 19°C (Figure 14). This data indicates that the palmitoyltransferase does not cycle through the Golgi or the TGN, and therefore its trafficking is not affected by this temperature block.

5. Discussion

The results presented in this study demonstrate that a membrane-bound enzyme is responsible for palmitoylation of the CD-MPR. Further characterization revealed that the palmitoyltransferase shuttles between the plasma membrane and endosomes.

Palmitoylation of a cysteine residue 34 amino acids distal from the trans-membrane domain is quite rare and the resulting membrane anchoring by the palmitate implies a drastic change in conformation of the entire cytoplasmic tail. The altered three-dimensional structure might have an effect on the exposure and accessibility of the sorting signals within the cytoplasmic tail of the receptor. To confirm the hypothesis that reversible palmitoylation of the CD-MPR regulates the sorting signals of the receptor, one prerequisite was to prove that an enzyme is involved in palmitoylation. Therefore we established an *in vitro* palmitoylation assay with purified CD-MPR and [^{3}H]palmitoyl-CoA as substrates and HeLa cell homogenate, containing the palmitoyltransferase activity. The *in vitro* assay uses the final substrates also required *in vivo* for palmitoylation. Hence the assay is very specific for the palmitoyltransferase activity and not dependent on additional enzymes such as the acyl-CoA synthase to synthesize [^{3}H]palmitoyl-CoA from [^{3}H]palmitate and coenzyme A. The CD-MPR, purified as a full length membrane protein from tissue culture cells is used in the assay, therefore containing the intact three-dimensional structure that might be essential for recognition by the palmitoyltransferase. The *in vitro* palmitoylation is abolished by boiling the HeLa cell homogenate to denature the palmitoyltransferase or by performing the assay on ice thereby inactivating the palmitoyltransferase. Furthermore, CD-MPR is not autocatalytically palmitoylated when incubated with [^{3}H]palmitoyl-CoA, thus demonstrating the requirement of an enzyme for palmitoylation. Palmitoylation is time- and temperature dependent, which are characteristics of an enzymatic reaction. All these results demonstrate that palmitoylation occurs enzymatically and further characterization revealed that the palmitoyltransferase is membrane-bound. The specificity of the palmitoyltransferase activity in the *in vitro* assay was further confirmed due to the fact that the lumenal cysteine residues were not palmitoylated in a mutant CD-MPR lacking the cytoplasmic cysteine residues (MPR-C30,34A).

To identify the intracellular localization of the palmitoyltransferase the cell homogenate was separated on a Percoll density gradient. A good separation of intracellular organelles was achieved for lysosomes, rough endoplasmic reticulum, the Golgi and the plasma membrane. The early and late endosomal compartments were dispersed over several fractions of the gradient, most likely due to the heterogeneous nature of these organelles. One peak of the marker for early endosomes was recovered in a very dense fraction (fraction 3) which might represent a coated/heavy subpopulation of an otherwise light organelle. A similar distribution was found for the late endosomes with a peak in a very dense fraction trailing into lighter fractions, also representing the heterogeneous nature of this organelle. The fractions were tested in the *in vitro* palmitoylation assay revealing that most of the total palmitoyltransferase activity was recovered in two peaks, one peak in a dense fraction and a second broad peak in light fractions. The same distribution was found for early endosomes. This could imply that the palmitoyltransferase is localized in early endosomes and therefore cofractionates with this organelle. Alternatively, it is also possible that the palmitoyltransferase is localized in more than one organelle and part of it fractionates with the corresponding organelle like early endosomes and the plasma membrane. To verify the localization of the palmitoyltransferase by additional methods, CD-MPR *in vivo* labeling experiments with [^{3}H]palmitate were performed, applying blocks to intracellular traffic to accumulate the receptor and possibly the palmitoyltransferase in a particular compartment. These experiments showed that the palmitoyltransferase cycles between the plasma membrane and endosomes. Wortmannin caused an accumulation of wt CD-MPR in the endosomes (Figure 12A, B) (Brown *et al.*, 1995; Kundra and Kornfeld, 1998). The increased level of palmitoylation of the wt CD-MPR upon addition of wortmannin demonstrated that the palmitoyltransferase is accumulated in endosomes as well. This accumulation was confirmed by the decreased extent of palmitoylation of the mutant CD-MPR at the plasma membrane upon treatment with wortmannin, showing that the palmitoyltransferase is depleted from the plasma membrane under this condition. Wortmannin does not affect the palmitoyltransferase reaction in general, which is demonstrated by both effects, a decrease and an increase, in the same experiment with different mutants. Thus wortmannin influences palmitoylation by changing the localization of the enzyme and the substrate indicating that the palmitoyltransferase cycles between the plasma membrane and endosomes. To determine whether the palmitoyltransferase cycles through the TGN as well, an incubation at low temperature (19°C) was used to specifically block vesicular transport out of the TGN (Matlin and Simons, 1983; Griffiths *et al.*, 1985). Thus, proteins like the wild-type CD-MPR, that cycle through the TGN as part of their normal trafficking route, are still delivered to the TGN, but fail to leave this organelle and therefore accumulate in the TGN upon reducing the temperature to 19°C. The mutant receptor is not affected by the temperature block since it is stuck at the plasma membrane and does not cycle at all. The decrease in palmitoylation of the wild-type CD-MPR upon the temperature block demonstrated that the palmitoyltransferase was not accumulated in the TGN together with wild-type receptor. In contrast to the wild-type receptor, both the mutant CD-MPR and the palmitoyltransferase were not affected by the temperature block and hence the level of palmitoylation of the mutant CD-MPR stayed unchanged. Furthermore, the normal level of palmitoylation of the mutant CD-MPR at 19°C was evidence to show that the incubation time and the kinetics of palmitoylation were sufficient to yield maximal extent of palmitoylation despite the lower temperature. The elevated level of palmitoylation of mutant receptors compared to wild-type receptors in non-treated cells reflects the different localizations of the receptors in relation to the palmitoyltransferase. At steady state wild-type receptors are mainly localized to the TGN, an organelle lacking the palmitoyltransferase activity, in contrast to mutant receptors that remain at the plasma membrane, which contains the palmitoyltransferase activity.

Altogether, we suggest a model where the palmitoyltransferase cycles between the plasma membrane and endosomes without passing through the TGN. This confirms the data from the Percoll gradient, where the palmitoyltransferase activity mostly colocalized with early endosomes and the plasma membrane.

Although we can not completely rule out a more complicated model involving two cytosolic thioesterases, one recruited to the TGN and one recruited to endosomes and the plasma membrane to explain our data, we consider this less likely. The increased and decreased levels of palmitoylation of the CD-MPR upon treatment with wortmannin (Figure 13) could be explained by a depletion of a thioesterase on endosomes (due to an effect of PI3K inhibition), which in turn has to induce an accumulation of this enzyme at the plasma membrane. The decrease in palmitoylation of the wild-type CD-MPR upon incubation at 19°C (Figure 14) would involve an accumulation of a thioesterase in the TGN, thus recruitment of a thioesterase to the TGN is required, in addition to endosomes and the plasma membrane. Upon accumulation at the TGN, this thioesterase should consequently be depleted from the other localizations, as it was the case upon treatment with wortmannin. However, the lack of a coinciding depletion of the thioesterase from the plasma membrane upon incubation at 19°C would indicate that two thioesterases are involved. Nevertheless, this model also requires palmitoyltransferase activity in both, endosomes and at the plasma membrane to account for the increased level of palmitoylation upon incubation with wortmannin (Figure 13) and the palmitoylation of the mutant CD-MPR, stuck at the plasma membrane (Figure 13 and Figure 14).

This leads us to the conclusion that a palmitoylation activity exists both at the plasma membrane and endosomes independent of which model holds true.

Palmitoyltransferase activities were reported in the early secretory pathway (Veit and Schmidt, 1993), in the Golgi (Gutierrez and Magee, 1991) and at the plasma membrane (Dunphy *et al.*, 1996). Our findings expand the knowledge about the localization of palmitoyltransferase activities with a transferase that cycles between endosomes and the plasma membrane. The variety of palmitoylated substrates and the lack of a clear consensus sequence suggest that several palmitoyltransferases with different specificities might exist. In mammals no candidate is known that could be the putative palmitoyltransferase of the CD-MPR. Two DHHC-CRD containing palmitoyltransferases identified in yeast, Akr1p and the Erf2p/Erf4p complex (Lobo *et al.*, 2002;

Roth et al., 2002), have different intracellular localizations and exhibit high substrate specificities, thereby validating the assumption of various palmitoyltransferases. Database searches revealed that there are 12 DHHC-CRD containing proteins in *Homo sapiens*, but none of them have been cloned so far. Further investigation will be required to determine if there is a cycling palmitoyltransferase among them, which cycles between the plasma membrane and endosomes with a specificity for the CD-MPR.

Furthermore, a palmitoyltransferase which cycles between the plasma membrane and endosomes could play a role in signal transduction. The G-protein-coupled receptor, β_2-adrenergic receptor (β_2AR), for example, is palmitoylated and this prevents phosphorylation of a nearby phosphorylation site. Upon activation by an agonist the β_2AR is depalmitoylated, followed by phosphorylation, desensitization and internalization. In endosomes β_2AR either gets targeted for degradation or dephosphorylated and subsequently recycled back to the plasma membrane with restoration of its native state by palmitoylation, possibly in the endosomes or at the plasma membrane (Pippig et al., 1995; Loisel et al., 1996; Moffett et al., 1996).

Moreover, endosomes are dynamic compartments displaying a highly complex and pleiomorphic organization (Gruenberg, 2001). No membrane protein has been reported so far that is restricted to endosomes, suggesting that a protein which is active in endosomes has to recycle, possibly via the plasma membrane.

Since the half-life of palmitoylation of the CD-MPR (less than 2 hours) is much less than the half-life of the CD-MPR (more than 40 hours) (Schweizer et al., 1996), it gets repeatedly palmitoylated during its life time. As a consequence, the palmitoyltransferase should be localized in a compartment that is part of the trafficking itinerary of the CD-MPR. The intracellular distribution of the CD-MPR comprises the TGN, the plasma membrane, early and late endosomes (Klumperman et al., 1993) and hence includes the organelles containing the palmitoyltransferase thereby enabling repeated palmitoylation of the CD-MPR. Moreover palmitoylation is essential for the CD-MPR in the late endosomes to avoid trafficking to the lysosomes (Schweizer et al., 1996). The localization of the palmitoyltransferase at the plasma membrane and endosomes is ideal to ensure that the CD-MPR is palmitoylated in the late endosomes thereby enabling its proper trafficking. The localization of the palmitoyltransferase to the site of its required function not only validates our results but also supports the hypothesis that palmitoylation might regulate the sorting of the CD-MPR by modulating the presentation of the sorting signals in the cytosolic tail.

6. Acknowledgements

We thank Drs. H.P. Hauri, J. Gruenberg, P. Chavrier and E. Berger for generously providing the antibodies. The hybridoma cell line producing mAb 22D4 against CD-MPR was a kind gift of Dr. D. Messner. P. Nair and B. Schaub are acknowledged for critical reading of the manuscript. We acknowledge Dr. E. Berger for continuous support and critical reading the manuscript. This work was supported in part by a Prof. Dr. Max Cloetta Fellowship to Jack Rohrer and Swiss National Science Foundation Grant 31-67274.

Part II:
The acidic cluster of the CD-MPR but not phosphorylation is required for GGA1 and AP1 binding

Jacqueline Stöckli, Stefan Höning and Jack Rohrer

1. SUMMARY 91

2. INTRODUCTION 92

3. MATERIALS AND METHODS 94

4. RESULTS 99

5. DISCUSSION 106

6. ACKNOWLEDGEMENT 109

1. Summary

Lysosomal biogenesis depends on proper transport of lysosomal enzymes by the cation-dependent mannose 6-phosphate receptor (CD-MPR) from the trans-Golgi network (TGN) to endosomes. Trafficking of the CD-MPR is mediated by sorting signals in its cytoplasmic tail. GGA1 (Golgi-localizing, γ-ear-containing, ARF-binding protein-1) binds to CD-MPR in the TGN, and targets the receptor to clathrin-coated pits for transport from the TGN to endosomes. The motif of the CD-MPR which interacts with GGA1 was shown to be D^{61}-X-X-L^{64}-L^{65}. Reports on increased affinity of cargo, when phosphorylated by casein kinase 2 (CK2), to GGAs, focused our interest on the effect of the CD-MPR CK2 site on binding to GGA1. Here we demonstrate that E^{58} and E^{59} of the CK2 site are essential for high affinity GGA1 binding *in vitro*, while the phosphorylation of S^{57} of the CD-MPR has no influence on receptor binding to GGA1. Furthermore, the *in vivo* interaction between GGA1 and CD-MPR was abolished only when all residues involved in GGA1 binding were mutated, namely, E^{58}, E^{59}, D^{61}, L^{64} and L^{65}. In contrast, binding of adaptor protein-1 (AP-1) to CD-MPR required all the glutamates surrounding the phosphorylation site, namely, E^{55}, E^{56}, E^{58} and E^{59}, but like GGA1 binding, was independent of the phosphorylation of S^{57}. The binding affinity of GGA1 to the CD-MPR was found to be 2.4-fold higher than that of AP-1. This could regulate binding of the two proteins to the partly overlapping sorting signals, allowing AP-1 binding to the CD-MPR only when GGA1 is released upon autoinhibition by phosphorylation.

2. Introduction

The cation-dependent mannose 6-phophate receptor (CD-MPR) is a type I integral membrane protein which is involved in the transport of lysosomal hydrolases to the lysosomes (Kornfeld, 1992; Hille-Rehfeld, 1995). Newly synthesized lysosomal enzymes acquire a mannose 6-phosphate (M6P) tag on their N-linked oligosaccharides by sequential action of two enzymes in the Golgi and the trans-Golgi network (TGN). (Bao et al., 1996b; Kornfeld et al., 1998). The M6P tag acts as lysosomal targeting signal and is recognized by the CD-MPR in the TGN. Upon binding lysosomal enzymes, CD-MPR is packaged into clathrin-coated vesicles and transported to the acidified endosomes, where the receptor dissociates from the ligand which is subsequently packaged into lysosomes. However, the receptor is transported either to the plasma membrane, where it is rapidly internalized, or recycled back to the TGN to mediate another round of sorting. The trafficking of the CD-MPR is directed by signals located in its 67 amino acid cytoplasmic tail.

Binding of the CD-MPR cytoplasmic tail to GGA1 (Golgi-localizing, γ-ear-containing, ARF-binding protein-1) mediates its transport out of the TGN (Puertollano et al., 2001a). GGA1 is a monomeric, soluble adaptor with four domains – an N-terminal VHS (Vps27p/Hrs/STAM) domain, a GAT (GGA and TOM1) domain, a connecting hinge segment and a C-terminal GAE (γ-adaptin ear) domain (Boman et al., 2000; Dell'Angelica et al., 2000; Hirst et al., 2000). Recruitment of GGA1 to the TGN is mediated by an interaction of the GAT domain with ADP ribosylation factor (ARF) (Collins et al., 2003). The VHS domain of GGA1 binds cargo which is subsequently targeted to clathrin-coated pits, mediated through an interaction between the GGA1 hinge domain with clathrin and adaptor protein-1 (AP-1) (Puertollano et al., 2001a; Doray et al., 2002b). The key residues in cargo, such as CD-MPR, cation-independent MPR (CI-MPR), sortilin, memapsin 2 and low density lipoprotein receptor-related protein 3 (LRP3), for binding to the VHS domain of GGA1 were shown to be DXXLL (Puertollano et al., 2001a; Takatsu et al., 2001; He et al., 2002; Misra et al., 2002). For the CI-MPR, it was shown that the phosphorylation by casein kinase 2 (CK2) of the serine preceding the aspartate increases its affinity to GGA1 (Misra et al., 2002). Furthermore, the acidic cluster-dileucine motif in the hinge domain of GGA1 requires phosphorylation by CK2 of the serine three residues upstream of the aspartate to facilitate autoinhibition of the VHS domain, by intramolecular or intermolecular binding (Doray et al., 2002a). The CD-MPR also contains a serine upstream of the D^{61}-X-X-L^{64}-L^{65}, which is phosphorylated by CK2, but its involvement in GGA1 binding has not been investigated so far (Hemer et al., 1993; Körner et al., 1994).

Reports on the functional importance of the CK2 phosphorylation site are controversial. Normal as well as impaired lysosomal enzyme delivery was reported for CD-MPR with mutant

CK2 phosphorylation sites (Johnson and Kornfeld, 1992b; Mauxion et al., 1996; Breuer et al., 1997). Furthermore, the phosphorylation of S^{57} was suggested to be required for surface delivery of the CD-MPR either directly from TGN or by inhibiting transport from endosomes to TGN (Breuer et al., 1997). On the other hand, phosphorylation of S^{57} was shown to be essential for AP-1 binding at the TGN for proper sorting to endosomes (Mauxion et al., 1996; Ghosh and Kornfeld, 2003a). However, Höning and colleagues (Höning et al., 1997) showed high affinity binding of AP-1 to non-phosphorylated CD-MPR peptides (residues 1-67 and 49-67). Altogether it is uncertain whether the phosphorylated serine or the non-phosphorylated serine leads to plasma membrane delivery of the CD-MPR, resulting in missorting of lysosomal enzymes, or whether it has an effect on lysosomal enzyme sorting at all.

Before the GGAs and their function of sorting in the TGN were identified, AP-1 was thought to mediate TGN sorting of cargo, such as CD-MPR (Ahle et al., 1988; Klumperman et al., 1993; Höning et al., 1997). Since the discovery of the GGAs the function of AP-1 became less clear and thereafter, some models accounting for the function of AP-1 were suggested. Fibroblasts deficient in µ1-subunit of AP-1 accumulated CD-MPR in endosomes and showed reduced retrograde transport *in vitro*, which led to the conclusion that AP-1 could be involved in retrograde transport from endosomes to the TGN (Meyer et al., 2000; Medigeshi and Schu, 2003). However, the localization of AP-1 to the TGN and in clathrin-coated vesicles originating from the TGN, as well as its increased binding to CD-MPR upon phosphorylation of the receptor led to a model where GGA1 recruits cargo, and by phosphorylation of both, GGA1 and cargo, GGA1 hands over the cargo to AP-1, which mediates vesicle budding from TGN (Klumperman et al., 1993; Doray et al., 2002b; Ghosh and Kornfeld, 2003a; Ghosh and Kornfeld, 2003b).

Our aim was to explore the involvement of the CK2 phosphorylation site in GGA1 and AP-1 binding in order to determine the importance of the CK2 phosphorylation of CD-MPR in TGN sorting. In this paper we show that two glutamates, E^{58} and E^{59} are involved in GGA1 binding *in vitro* in addition to the known DXXLL motif and that phosphorylation of S^{57} of the CD-MPR had no effect on GGA1 binding. The importance of E^{58} and E^{59} of the CD-MPR for high affinity binding of GGA1 was further confirmed by kinetic studies. Analysis of the interaction of GGA1 with the CD-MPR *in vivo* demonstrates that the binding was abolished, only when residues E^{58}, E^{59}, D^{61}, L^{64} and L^{65} were mutated simultaneously. A comparison of the binding properties of AP-1 and GGA1 to the CD-MPR revealed subtle but essential differences. Most importantly the CD-MPR has a lower binding affinity to AP-1 than to GGA1. Furthermore, all four glutamates, E^{55}, E^{56}, E^{58} and E^{59}, but not the phosphorylation of S^{57}, were involved in AP-1 binding.

3. Materials and Methods

Materials

Enzymes used in molecular cloning were obtained from Roche Diagnostics (Mannheim, Germany), New England Biolabs (Beverly, MA, USA), or Promega (Madison, WI, USA); general chemicals from Fluka (Buchs, Switzerland); protease inhibitors from Sigma (St. Louis, MO, USA); Dulbecco's Modified Eagle Medium (DMEM), fetal calf serum (FCS), G418 and Lipofectamine Plus were from Invitrogen (Carlsbad, CA, USA); polyethylenimine, 25 kDa (CAT# 23966) from Polysciences, Inc. (Warrington, PA, USA); cell culture dishes from Falcon (Franklin Lakes, NJ, USA); nitrocellulose from Schleicher & Schuell (Dassel, Germany); enhanced chemiluminescence Western blotting reagents from PerkinElmer Life Sciences (Boston, MA, USA); Glutathione Sepharose 4B and low molecular weight protein markers from Amersham Pharmacia Biotech (Piscataway, NJ, USA); Prolong Antifade from Molecular Probes (Eugene, OR, USA). Oligonucleotides were synthesized either by the DNA synthesis facility of the Friedrich Miescher Institute (Basel, Switzerland) or Microsynth GmBH (Balgach, Switzerland).

Antibodies

Horseradish peroxidase conjugated antibodies against mouse and rabbit were from Amersham Pharmacia Biotech (Piscataway, NJ, USA). Alexa 568 conjugated goat anti-mouse antibody was from Molecular Probes (Eugene, OR, USA). The monoclonal antibody 22D4 specific for the bovine CD-MPR was generously provided by D. Messner (Messner, 1993). This monoclonal antibody is specific for the bovine CD-MPR and does not cross-react with the endogenous mouse CD-MPR (Rohrer and Kornfeld, 2001).

Recombinant DNA

All basic DNA procedures were as described (Sambrook et al., 1998). The PCR procedure of Ho and colleagues (Ho et al., 1989) was used to generate the MPR-FFWYLL-A, MPR-$C^{30,34}$A, MPR-S^{57}A, MPR-S^{57}D, MPR-Clus⁻S, MPR-Clus⁻A, MPR-Clus⁻D, MPR-E^{55}Q, MPR-E^{56}Q, MPR-$E^{55,56}$Q, MPR-E^{58}Q, MPR-E^{59}Q, MPR-$E^{58,59}$Q, MPR-D^{61}N, MPR-D^{61}N,$L^{64,65}$A, MPR-$E^{58,59}$Q,D^{61}N,$L^{64,65}$A constructs with pSFFV-MPR (Rohrer et al., 1995) serving as a template together with MPR-BglII.down (5'-CCGAGATCTCCCACTTAAGCGTGG-3') and pSFFVneo.up2 (5'-CTGCCATTCATCCGCTTATTATC-3') as the down- and upstream primers respectively. Appropriate partial complementary pairs of oligonucleotides in which the desired amino acid replacement had been incorporated were chosen as internal primers. The final PCR products were subcloned into pSFFVneo as described (Rohrer et al., 1995) and confirmed by sequencing.

The GST-GGA1-A240stop construct was generated by PCR-amplification, using myc-GGA1pFB1 (a generous gift from Stuart Kornfeld, Washington University School of Medicine, St. Louis, MO, USA) as a template and subsequent cloning into the BamHI-NotI sites of pGEX-6p3 (Amersham Pharmacia Biotech, Piscataway, NJ, USA). The GFP-GGA1-A240stop construct was generated in two steps as follows. The GGA1-A240stop insert was subcloned into the BamHI-NotI sites of pcDNA3.1+ which was then digested with XhoI, the overhanging 5' end was filled up with the Klenow fragment of DNA polymerase I and subsequently digested with KpnI. The fragment was ligated with KpnI-SmaI-digested pEGFP-C1 vector (BD Biosciences Clonetech, Palo Alto, CA, USA). The construct was confirmed by sequencing.

Cell Culture and Transfection

A mannose-6-P/insulin-like growth factor-II receptor-deficient mouse L cell line designated D9 (LRec−) was maintained in DMEM containing 10% FCS. The cells were transfected with $XbaI$-linearized DNA with Lipofectamine Plus according to the manufacturer's directions. Selection for resistance to neomycin (G418) was carried out using 500 µg/ml G418 as the final concentration. Resistant colonies were picked individually and screened for the expression of bovine CD-MPR by immunoblotting. Clones expressing similar amounts of receptor compared to ML4 cells, the reference cell line expressing wt bovine CD-MPR (Johnson et al., 1990), were expanded for further study and maintained in selective medium.

Mouse L cells stably expressing wild-type or mutant CD-MPR were grown on coverslips in 6-well plates in DMEM to 50% confluency before transient transfection with pEGFP-GGA1-A240stop using polyethylenimine (25 kDa). 2.2 µg DNA in 75 µl DMEM was mixed with 8 µl 1 mg/ml polyethylenimine in 75 µl DMEM, vortexed and incubated for 15 min at room temperature following addition of 1.05 ml DMEM. The mix was added to the cells in the 6-well plate and after 24 h the cells were fixed and prepared for immunofluorescence.

In Vitro GST Pulldown Experiment

GST-GGA1-A240stop or GST alone was expressed in *Escherichia coli* strain DH5α. A saturated overnight culture was diluted 1:10 in 25 ml of growth medium and incubated at 37°C until OD_{600} was 0.6-0.8 before induction with 1 mM isopropyl-1-thio-β-D-galactopyranoside (IPTG) for 3 h. The cells were harvested by centrifugation, washed with ice-cold phosphate-buffered saline (PBS), and lysed by sonication in 2 ml pulldown buffer (50 mM Hepes, pH 7.4, 150 mM KCl, 1 mM $MgCl_2$) containing a 1:500 dilution of a protease inhibitor cocktail (5 mg/ml benzamidine, and 1 mg/ml each of pepstatin A, leupeptin, antipain, and chymostatin in 40% dimethyl sulfoxide, 60% ethanol) (PIC) and phenylmethylsulfonyl fluoride (PMSF) (40 µg/ml in ethanol). Insoluble material was removed by centrifugation at 12'000 rpm for 10 min at 4°C in a Sorvall GSA centrifuge. The supernatant was incubated for 30 min at room temperature on a rotating shaker with 400 µl Glutathione Sepharose 4B beads which were prewashed three times with pulldown buffer containing 0.1% bovine serum albumin in a silanized Eppendorf tube. The beads with GST or GST-GGA1-A240stop were washed three times and then resuspended in 1 ml pulldown buffer containing PIC and PMSF, being the amount for ten assays.

Extracts from mouse L cells expressing wild-type or mutant CD-MPR constructs were prepared from cells grown on 15-cm Falcon tissue culture dishes in the following way. Confluent 15-cm dishes of cells were put on ice and washed once with 10 ml of ice-cold PBS and scraped in 5 ml of pulldown buffer containing PIC and PMSF. The cells were pelleted at 1000 rpm for 5 min at 4°C in a Heraeus centrifuge and resuspended in 1 ml of pulldown buffer containing PIC and PMSF. The cells were homogenized in a ball-bearing homogenizer of 16 µm clearance using 12 strokes on ice and centrifuged at 700xg for 10 min at 4°C. The protein concentration of the resulting post-nuclear supernatant (PNS) was measured using a protein assay (Bio-Rad, Hercules, CA, USA). The appropriate amounts of PNS from each of the mutant cell lines and wild-type cell line yielding the same amount of CD-MPR were determined by Western blotting with anti-CD-MPR monoclonal antibody 22D4 followed by densitometric scanning and quantification using ImageQuant 5.0 software (Amersham Pharmacia Biotech, Piscataway, NJ, USA). These PNS samples adjusted for expression levels were used for the assay.

100 µl of the resuspended beads bound with either GST or GST-GGA1-A240stop were incubated with 100 µg of PNS from the wild-type cell line or an equivalent amount of PNS from mutant cell lines adjusted for the expression level as described above in 300 µl pulldown buffer containing PIC and PMSF and 0.1% Triton X-100 in a silanized Eppendorf tube for 2 h at 4°C on a rotating shaker. Beads were spun at 2500 rpm for 2 min at room temperature in an Eppendorf tabletop centrifuge. The supernatant was collected and stored. The beads were washed three times with pulldown buffer containing 0.1% Triton X-100. 40 µl non-reducing SDS-PAGE sample buffer (94 mM Tris-HCl (pH 6.8), 3% SDS, 15% glycerol, 0.001% bromophenol blue) was added to the beads, boiled and analyzed by SDS-PAGE and Western blotting using anti-CD-MPR monoclonal antibody 22D4. An aliquot of each of the supernatants was also analyzed similarly to determine the amount of unbound receptor.

Purification of GST-GGA1-A240stop

GST-GGA1-A240stop was bound to Glutathione Sepharose 4B beads as described above. The procedure was scaled up to 400 ml bacterial overnight culture, which were diluted to 4 L. The volumes of the following steps were adjusted as follows; the bacteria were sonicated in 90 ml pulldown buffer and the amount of Glutathione Sepharose 4B was 1 ml. After washing the beads three times with pulldown buffer, the GST fusion protein was eluted by incubating the beads with 1 ml elution buffer (50 mM Tris pH 8.0, 10 mM glutathione reduced, 2 mM DTT) containing PIC and PMSF for 30 min at room temperature. The beads were spun down at 700 x g for 2 min and the supernatant containing the GST-GGA1-A240 protein was dialysed against buffer A (50 mM Tris pH 8.7, 250 mM NaCl, 1 mM DTT) and GST-GGA1-A240 was recovered and the concentration was measured using a protein assay.

Analysis of Protein-Protein Interaction by SPR

To monitor binding of AP-1 and GGA1 to the CD-MPR carboxy-terminal domain, a synthetic peptide corresponding to the tail residues 49-67 and mutants within this peptide were immobilized on a CM5 surface of a BIAcore 3000 biosensor. AP-1 was purified form pig brain using published methods (Höning et al., 1997), GGA1 was purified as a GST-fusion protein. Both proteins were used at concentrations ranging from 25 – 500 nM in buffer A (50 mM Tris pH 8.7, 250 mM NaCl, 1 mM DTT) at a flow-rate of 20 µl/min. Binding and dissociation were recorded for 2 min. A short pulse injection (5 sec) of 50 mM NaOH was then used to remove bound material from the sensor surface. The rate-constants were determined using the software supplied by the manufacturer assuming a single binding site for AP-1 and GGA1.

SDS-PAGE and Immunoblotting

The proteins were separated on a 10% SDS-polyacrylamide minigel by using the Laemmli system (Laemmli, 1970). After electrophoresis, gels were transferred onto nitrocellulose membranes according to the method of Towbin (Towbin et al., 1979). The membrane was blocked with 3% nonfat dry milk powder (Sano Lait, Coop, Switzerland) in PBS. The blot was subsequently incubated with mAb 22D4 (diluted 1:500 in PBS-3% powdered milk) followed by a horseradish peroxidase conjugated anti-mouse secondary antibody (diluted 1:2000 in PBS-3% powdered milk). Immunoreactive proteins were visualized using the enhanced chemiluminescence detection system according to the manufacturer's directions.

Confocal Immunofluorescence Microscopy

Cells were grown on coverslips and transiently transfected with pEGFP-GGA1-A240stop. After 24 h, cells were washed with PBS and fixed in 3% paraformaldehyde pH 8.3 for 20 min, followed by four washes with 20 mM glycine in PBS. The cells were permeablized in saponin buffer (0.1% saponin, 20 mM glycine in PBS) for 20 min. All following steps were performed in saponin buffer. Cells were incubated with anti-CD-MPR monoclonal antibody 22D4 (1:500) for 30 min and washed four times followed by the incubation with goat anti-mouse Alexa 568 antibody. The coverslips were washed four times and mounted on glass slides with ProLong Antifade for viewing with a Leica SP2 AOBS UV confocal laser-scanning microscope. Serial sections in the z axis through the entire cells were taken, and the resulting stacks of images were analyzed using the Imaris program (Bitplane AG, Zürich, Switzerland).

4. Results

Acidic cluster of the CK2 site of CD-MPR but not phosphorylation is essential for GGA1 binding

In order to analyze the involvement of the CK2 phosphorylation site of the CD-MPR in GGA1 binding, several mutants of the bovine CD-MPR were constructed and stably transfected into mouse L cells. We used full length CD-MPR, expressed in mammalian cells to allow the formation of the correct post-translational modification and the three-dimensional structure, which provided physiological conditions for protein-protein interactions. To differentiate between the phosphorylation and the acidic CK2 site, several point mutations of the CD-MPR sequence were

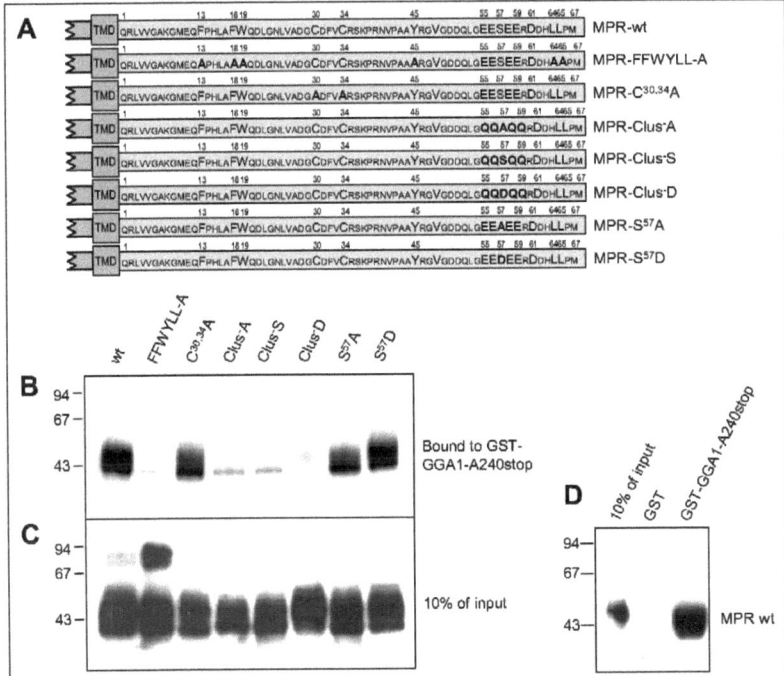

Figure 15: Interaction *in vitro* between wild-type and mutant CD-MPR constructs with GGA1. *A*, Schematic illustration of the cytoplasmic tails of CD-MPR wildtype and mutants. The 67 amino acids of the cytoplasmic tail are shown in single letter code. In the mutant constructs the mutated amino acids are indicated by bold letters. *B*, GST-GGA1-A240stop pulldown with CD-MPR wild-type and mutants. GST-GGA1-A240stop was purified from bacterial culture and bound to Glutathione Sepharose 4B beads, which was subsequently incubated with post-nuclear supernatant (PNS) from mouse L cells, stably transfected with CD-MPR wild-type or mutants. Following incubation at 4°C for 2 h the Sepharose was washed three times, boiled in sample buffer und subjected to SDS-PAGE. Bound receptor was detected by immunoblotting with anti-CD-MPR antibody. *C*, To detect CD-MPR expression levels, 10% of the PNS used for the GST-GGA1-A240stop pulldown were subjected to SDS-PAGE and immunoblotting with anti-CD-MPR antibody. *D*, Pulldown of GST and GST-GGA1-A240stop with PNS from mouse L cells stably transfected with wild-type CD-MPR. Protein standard shows molecular mass in kDa.

created (Figure 15A). The S^{57} was either replaced by an alanine to disable phosphorylation or by an aspartate to disable phosphorylation, but mimicking the negative charge of phosphorylation. In addition to the three variants at position 57, the cluster of four glutamates ($E^{55,56,58,59}$) of the CK2 site were mutated to glutamines, generating the cluster-minus mutants (MPR-Clus¯S, MPR-Clus¯A and MPR-Clus¯D). Apart from the CK2 site mutants, two additional mutants were tested for GGA1 binding, the palmitoylation-deficient (MPR-$C^{30,34}$A) mutant and the internalization-deficient (MPR-FFWYLL-A) mutant, lacking the dileucine motif, which is part of the known GGA1 binding motif (D^{61}-X-X-L^{64}-L^{65}) (Puertollano et al., 2001a; Takatsu et al., 2001; Misra et al., 2002). The GGA1 binding assay was performed with wild-type or mutant CD-MPR, in post-nuclear supernatant (PNS) from stably transfected cells and a truncated form of GGA1 fused to GST (GST-GGA1-A240stop), since the full-length GGA1 has been shown to be autoinhibitory for binding to the receptor (Doray et al., 2002a). GST-GGA1-A240stop comprises the VHS domain for cargo binding and a larger part of the GAT domain for the interaction with ARF. In the GST pulldown assay we could detect strong binding of the wild-type MPR to the GGA1 fragment and no binding of the FFWYLL-A mutant, as expected (Figure 15B), with the PNS containing an equal amount of the CD-MPR protein (Figure 15C). The specificity of the assay was confirmed by the fact that the CD-MPR did bind to the GST-GGA1-A240stop and not to GST alone (Figure 15D). The palmitoylation-deficient CD-MPR mutant (MPR-$C^{30,34}$A) bound to GGA1 comparable to wild-type (Figure 15B,C). This mutant is known to accumulate in lysosomes due to the lack of palmitoylation (Schweizer et al., 1996), however the deficiency in palmitoylation didn't influence the recognition of GGA1, suggesting that this posttranslational modification is not essential for TGN sorting. Interestingly, the binding of the S^{57} mutants (MPR-S^{57}A, MPR-S^{57}D) to GGA1 was not impaired, indicating that phosphorylation doesn't play a role in GGA1 binding (Figure 15B,C). However, the cluster-minus mutants, in which the negatively charged glutamates were changed to non-charged glutamines, did not bind to GGA1. This result was again independent of the amino acid substitution at position 57.

E^{58} and E^{59} of the CK2 site are essential for GGA1 binding

In order to identify more precisely the residues of the CK2 site in the CD-MPR which are responsible for the impaired binding to GGA1, we created mutant receptors with single and double exchanges of the glutamates ($E^{55,56,58,59}$) (Figure 16A). Additionally, the residues composing the motif already known to be essential for GGA1 binding were replaced to generate the mutants MPR-D^{61}N and MPR-D^{61}N,$L^{64,65}$A (Puertollano et al., 2001a; Takatsu et al., 2001; Misra et al., 2002). All mutant CD-MPR constructs were stably transfected into mouse L cells and the PNS of these cells was used in the GGA1 binding assay. Changing the charged residues E^{55} and E^{56} of the cytoplasmic tail to glutamines did not have an effect on GGA1 binding, neither as single mutants (MPR-E^{55}Q,

MPR-E^{56}Q) nor combined as a double mutant (MPR-$E^{55,56}$Q) (Figure 16B,C). However, binding of MPR-E^{58}Q to GGA1 was reduced and binding of MPR-E^{59}Q was affected even more. The binding of MPR-$E^{58,59}$Q to GGA1 was not detectable, like the CD-MPR mutants known for impaired GGA1 binding, such as MPR-D^{61}N and MPR-D^{61}N,$L^{64,65}$A (Figure 16B,C). Thus, the complete GGA1 binding motif of the CD-MPR includes E^{59} and to some extend E^{58}: E^{58}-E^{59}-X-D^{61}-X-X-L^{64}-L^{65}.

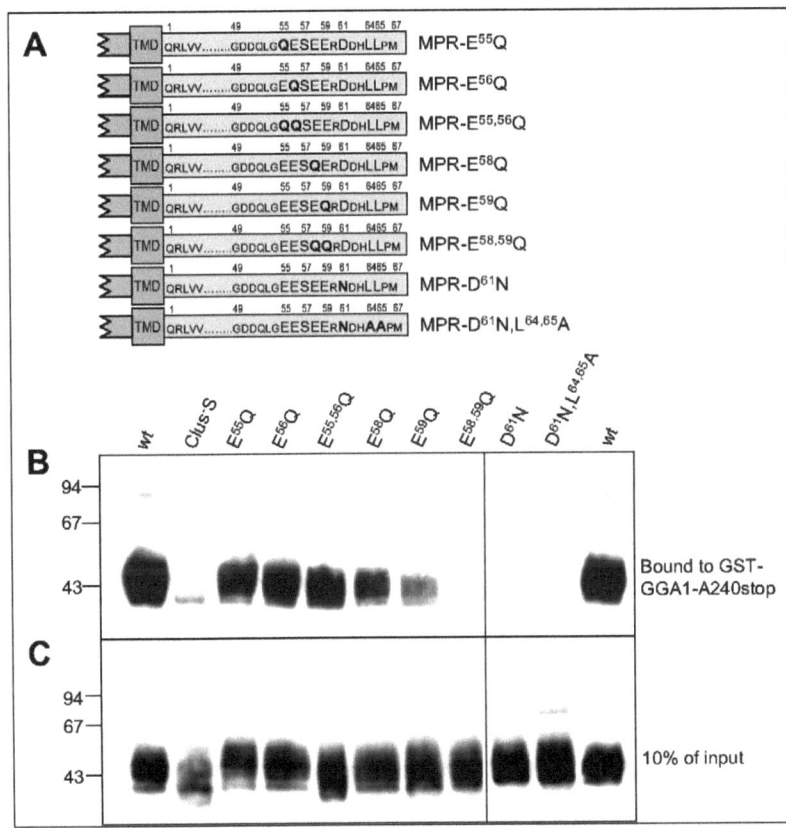

Figure 16: Binding analysis of casein kinase 2 site of the CD-MPR with GGA1. *A*, Schematic illustration of the cytoplasmic tails of CD-MPR wildtype and mutants. The amino acids of the cytoplasmic tail are shown in single letter code. In the mutant constructs the mutated amino acids are indicated by bold letters. *B*, GST-GGA1-A240stop pulldown with CD-MPR wild-type and mutants. GST-GGA1-A240stop was purified from bacterial culture and bound to Glutathione Sepharose 4B, which was subsequently incubated with post-nuclear supernatant (PNS) from mouse L cells, stably transfected with CD-MPR wild-type or mutants. Following incubation at 4°C for 2 h the Sepharose was washed three times, boiled in sample buffer und subjected to SDS-PAGE. Bound receptor was detected by immunoblotting with anti-CD-MPR antibody. *C*, To detect CD-MPR expression levels, 10% of the PNS used for the GST-GGA1-A240stop pulldown was subjected to SDS-PAGE and immunoblotting with anti-CD-MPR antibody. Protein standard shows molecular mass in kDa.

Mutation of E^{58} and E^{59} to alanines increased rate-constants for GGA1 binding

To analyze the kinetics of the GGA1 - CD-MPR interaction and to verify the results obtained in the GST-GGA1-A240 pulldown experiments, surface plasmon resonance (SPR) technology was used. For this purpose CD-MPR tail peptides from residue 49 to 67, with certain residues replaced by alanines (Figure 17A) and either containing a normal serine or a phospho-serine, were coupled to the surface of a biosensor. GGA1 was purified as GST-GGA1-A240stop fusion protein and used at concentrations ranging from 25-500 nM. Binding and dissociation were recorded for 2 min and the rate-constant was calculated. The CD-MPR wild-type peptide and the MPR-$E^{55,56}$A peptide both bound to GGA1 with a comparable rate-constant (K_D). For both peptides phosphorylation had only a minor effect, decreasing the K_D by 0.17-fold (Figure 17B). All the other mutations in the peptides caused an increased K_D, indicating their requirement for GGA1 binding. The affinity of the binding of GGA1 to the $D^{61,62}$A peptide was too low to be detected. The $L^{64,65}$A peptide showed a 6-fold increase in K_D compared to the CD-MPR wild-type peptide and its phosphorylation increased binding of GGA1 only by 0.18-fold. The new residues found to be essential for GGA1 – CD-MPR interaction in the GST pulldown experiment ($E^{58,59}$) were also required for binding in the SPR experiment. Compared to MPR-wt peptide the MPR-$E^{58,59}$A peptide showed a 3.2-fold increased rate-constant for GGA1 binding. Furthermore, this peptide displayed a major difference between the non-phosphorylated and the phosphorylated form. The GGA1 had almost a normal binding affinity to the phosphorylated MPR-$E^{58,59}$A peptide, but not to the non-phosphorylated form, thus a 2.6-fold difference in rate-constant was measured. Hence, the binding of GGA1 to the MPR-$E^{58,59}$A peptide could be restored by the phosphorylation of S^{57}, whereas the phosphorylation in the other mutants displayed only a minor increase in binding of GGA1. The inhibition of GGA1 binding was tested by incubating GGA1 with short inhibitory peptides prior to the binding assay. GGA1 was incubated with a 10-fold molar excess of soluble peptide A (49-58) or B (58-67) (Figure 17A) for 15 min, followed by recording of the binding to the immobilized MPR wild-type peptide on the biosensor. Only peptide B was able to block binding to the MPR wild-type peptide, while peptide A had no effect (data not shown), confirming the data from the mutational analysis.

Acidic cluster of the CK2 site of CD-MPR but not phosphorylation is essential for AP-1 binding

In order to explore the involvement of the CK2 phosphorylation site of the CD-MPR in AP-1 binding and its kinetics, SPR experiments were performed. The same peptides as for the GGA1 binding were used (Figure 17A). Purified AP-1 was used at concentrations ranging from 25-500 nM. Binding and dissociation were recorded for 2 min and the rate-constant was calculated. The

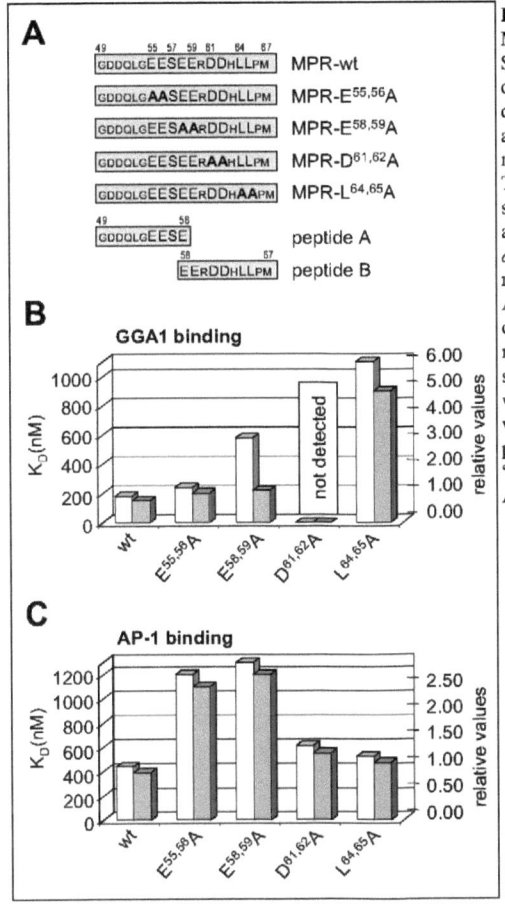

Figure 17: Binding of GGA1 and AP-1 to CD-MPR wild-type and mutant peptides. *A*, Schematic illustration of the peptides, corresponding to the amino acids 49 to 67 of the cytoplasmic tail of the CD-MPR. The amino acids are shown in single letter code and the mutated amino acids are indicated by bold letters. The five upper peptides were coupled to the sensor chip surface, the bottom two peptides, A and B, were used to test inhibition of binding. *B and C*, Binding assays with surface plasmon resonance technology. Binding of GST-GGA1-A240stop (*B*) and AP-1 (*C*) to the peptides either containing a phospho-serine (*gray columns*) or a non-phosphorylated serine (*white columns*) was subsequently recorded and the rate-constants were calculated (*left axis*). The relative values were calculated compared to the non-phosphorylated wild-type peptide (*right axis*). *B*, "not detected" indicates that GST-GGA1-A240stop did not bind to the $D^{61,62}$A peptide.

affinity of AP-1 to the wild-type CD-MPR peptide was 2.4-fold lower than that of the GGA1 to the same peptide (Figure 17B,C). This result explained the difficulties in detecting AP-1 in a co-immunoprecipitation of CD-MPR or GST-pulldown experiments (data not shown). For binding of AP-1 to the peptides MPR-$D^{61,62}$A and MPR-$L^{64,65}$A the rate-constants were in a similar range as for CD-MPR wild-type peptide (Figure 17C). However, the mutation of both pairs of glutamates to alanines, MPR-$E^{55,56}$A and MPR-$E^{58,59}$A, increased the rate-constant by a factor of 2.7 and 3.0, respectively (Figure 17C). In all peptides, phosphorylation of the serine only showed a minor increase in binding to AP-1, 0.1-fold. Thus, the glutamates of the CK2 phosphorylation site, but not the phosphorylation itself are essential for AP-1 binding.

TGN redistribution of CD-MPR in vivo by dominant-negative GGA1 is impaired in the mutant containing $E^{58,59}Q, D^{61}N, L^{64,65}A$ mutations

In order to determine whether the *in vitro* results reflect the situation in living cells, interaction between GGA1 and CD-MPR was investigated *in vivo*. It was previously shown that a dominant-negative GGA1 containing the VHS domain for cargo binding and the GAT domain to be recruited to the TGN could redistribute and accumulate CI-MPR and CD-MPR in the TGN (Puertollano et al., 2001a; Puertollano et al., 2001b). GFP-GGA1-A240stop, although not containing the complete GAT domain, comprised the essential domains for interaction with ARF to

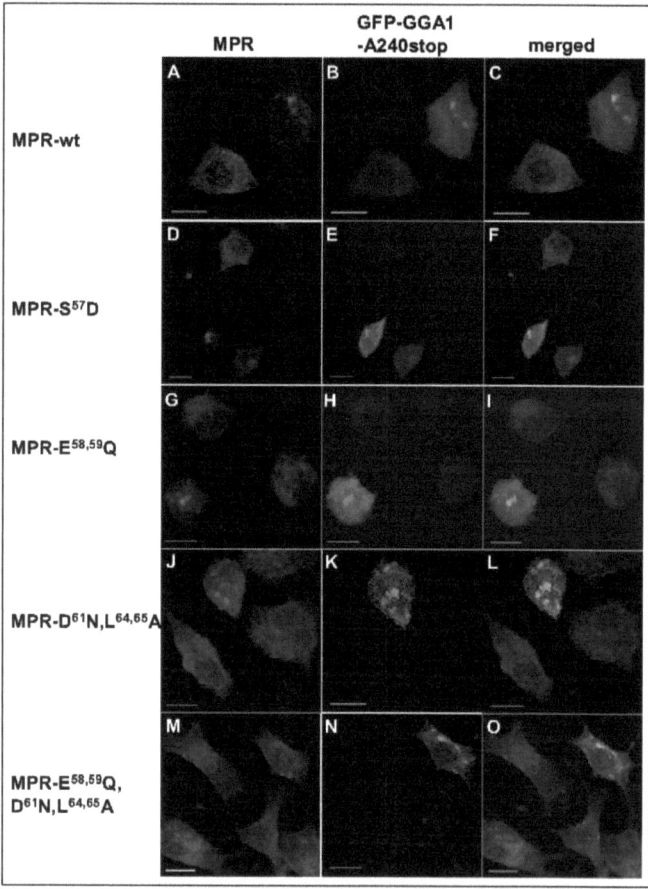

Figure 18: Interaction of dominant-negative GGA1 with CD-MPR wild-type and mutants. A-O, Mouse L cells stably transfected with CD-MPR wild-type (A-C) and mutants (D-O) were grown on coverslips. Cells were transiently transfected with dominant-negative GFP-GGA1 (green), fixed and permeabilized after 24 h. Cells were incubated with anti-CD-MPR monoclonal antibody 22D4, followed by incubation with goat anti-mouse Alexa 568 antibody (red). Serial sections in the z axis through the entire cells were taken with a confocal laser-scanning microscope. Scale bars are 10 µm.

be recruited to the TGN, but lacked the hinge and GAE domains and is therefore unable to recruit clathrin and accessory proteins required for budding (Collins et al., 2003). Thus, GFP-GGA1-A240stop was acting as a dominant-negative GGA1 (Puertollano et al., 2001a), when transiently transfected into mouse L cells, stably transfected with CD-MPR wild-type (Figure 18A-C). The transfected cells showed a redistribution of the CD-MPR to the TGN and consequently a depletion from the periphery. As expected, the CD-MPR mutants, which did interact with GGA1 *in vitro*, were redistributed to the TGN as well, when dominant-negative GGA1 was expressed (Figure 18D-F and data not shown). Interestingly, the CD-MPR mutants MPR-$D^{61}N,L^{64,65}A$ and MPR-$E^{58,59}Q$ presented the same effect indicating that their binding to GGA1 was not completely impaired, although they were not detectable in the GST-GGA1 pulldown and the corresponding peptides had a very high K_D for the interaction with GGA1 (Figure 18G-L). Therefore another mutant construct was created, where all residues involved in GGA1 binding were changed simultaneously, the MPR-$E^{58,59}Q,D^{61}N,L^{64,65}A$. The transient transfection of dominant-negative GGA1 did not redistribute the MPR-$E^{58,59}Q,D^{61}N,L^{64,65}A$ to the TGN, indicating that GGA1 is not interacting with this mutant CD-MPR (Figure 18M-O).

5. Discussion

The results presented in this study demonstrate that the acidic cluster of the CK2 site upstream of the DXXLL motif of the CD-MPR but not its phosphorylation is required for GGA1 and AP-1 binding. Furthermore we show that only the CD-MPR mutant with all residues of the GGA1 binding site changed (MPR-E^{58,59}Q,D^{61}N,L^{64,65}A) is unable to interact with GGA1 *in vivo*. The different affinities of CD-MPR to AP-1 and GGA1 might present a regulatory mechanism of the binding of these adaptors to the overlapping sorting signals in the cytoplasmic tail of the CD-MPR.

GGAs have been shown to be essential for transport of CD-MPR and other cargo from the TGN to endosomes. The consensus binding motif of the VHS domains of GGAs and their cargo was revealed to be DXXLL (Puertollano et al., 2001a; Takatsu et al., 2001; Zhu et al., 2001). Given that the CK2 phosphorylation of a serine upstream of the DXXLL motif increased binding affinity of the CI-MPR peptide to the VHS domains of GGA1 and GGA3, additional residues might be involved in GGA binding (Kato et al., 2002). Furthermore, the autoinhibitory function of the acidic cluster-dileucine motif in the hinge domain of GGA1 and GGA3 by intramolecular or intermolecular binding of the VHS domain depends on the CK2 phosphorylation of the serine three residues upstream (Doray et al., 2002a). The existence of a CK2 phosphorylation site upstream of the DXXLL motif in the CD-MPR suggested a similar influence on GGA1 binding. However, our results revealed that the phosphorylation of S^{57} had no effect on GGA1 binding. Mutants with the serine replaced by an alanine or an aspartate to mimic phosphorylation bound GGA1 similar to wild-type receptor. Conversely, the mutations of the glutamates surrounding the serine in the receptor led to an inhibition of GGA1 binding. Further characterization of the glutamates revealed the E^{58} and E^{59} residues to be essential for GGA1 binding. The kinetic studies of the interaction between GGA1 and CD-MPR confirmed that indeed the MPR-E^{58,59}A peptide showed a 3-fold increased rate-constant compared to wild-type. The rate-constants for the residues of the DXXLL motif were even higher, the K$_D$ of the MPR-L^{64,65}A peptide being 6-fold increased, and the K$_D$ of the MPR-D^{61,62}A peptide being not measurable, reflecting the very low affinity of GGA1 to these residues. The *in vivo* interaction of CD-MPR with dominant-negative GGA1 demonstrated that mutating the residues D^{61}N,L^{64,65}A and the E^{58,59}Q, individually, did not inhibit GGA1 binding. Both mutant constructs redistributed and accumulated in the TGN upon expression of the dominant-negative GGA1, showing a clearly different pattern compared to non-transfected cells. Therefore the CD-MPR containing all amino acids mutated, namely the new residues from this study, E^{58,59}Q, combined with the already known residues, D^{61}N,L^{64,65}A, was generated. Upon expression of dominant-negative GGA1, no redistribution and accumulation of the MPR-E^{58,59}Q,D^{61}N,L^{64,65}A was observed. This indicates that the interaction of MPR-E^{58,59}Q,D^{61}N,L^{64,65}A with GGA1 is inhibited. Altogether, these results led to the extension of the GGA1 interaction motif of the CD-MPR to E^{58}-E^{59}-X-D^{61}-X-X-L^{64}-L^{65}.

The alignment of the GGA1 binding sites of cargo revealed that all the proteins have in the positions upstream from the D^{61} up to the E^{58} in the CD-MPR (position -3 to -1 from the crucial aspartate 0, Figure 19) at least one acidic residue or a serine phosphorylation site (Figure 19, gray box, italic letters). This indicates that a negative charge, not specifically a glutamate, within this range of the position -1 to -3 might be crucial for GGA1 binding. This is confirmed by the requirement of the phosphorylation of the serine at position -3 in the GGA1 hinge domain for its autoinhibitory function and the increased affinity to GGA when the serine at position -1 in the CI-MPR is phosphorylated (Figure 19).

```
               -4    0     4
CD-MPR:     -GEESEERDDHLLPM-COOH
CI-MPR:     -VSFHDDSDEDLLHI-COOH
Sortilin:   -SGYHDDSDEDLLE-COOH
LRP3:       -PMLEASDDWALLVC-COOH
Memapsin2:  -QHDDFADDISLLK-COOH
GGA1-hinge: -SASVSLLDDFLMSLGL-
```

Figure 19: GGA binding domain in cargo proteins. C-terminal sequences of CD-MPR, CI-MPR, sortilin and LDL receptor related protein 3 (LRP3) and the internal autoinhibitory motif in the GGA1 hinge domain are shown in the amino acid single letter code. The amino acids involved in binding to the VHS domain of GGA are indicated by bold letters. Residues are numbered relative to Asp 0. The gray box specifies the region where negative charges were shown to be essential for GGA binding. Negatively charged residues or phosphorylation sites, which are located in the gray box and could be involved in GGA1 binding, are indicated with italic letters.

(Doray et al., 2002a; Kato et al., 2002). Furthermore, the rescue of high affinity GGA1 binding in the $E^{58,59}$A peptide, when the serine was phosphorylated (Figure 17B), led to the assumption, that the required negative charges for GGA1 binding could be compensated by enough negative charges in the vicinity, which might range maximally between position -4 to -1 (Figure 19). The data from the structural analysis with VHS domains of GGA and CI-MPR or CD-MPR peptides revealed that the residues -6 to -3 were disordered and the residues from -2 to 0 were well ordered with the aspartate 0 forming the most extensive interaction with the VHS domain (Misra et al., 2002; Shiba et al., 2002). Three positively charged amino acids in the GGA1 and GGA3 VHS domains interact with the cargo peptide. K^{131} of GGA1 (GGA3 amino acid numbers are decreased by one) mainly interacts with D at position 0, whereas K^{87} and R^{89} interact with phospho-serine at position -1 of the CI-MPR (Kato et al., 2002; Shiba et al., 2002). It might be possible that the latter two basic residues of the VHS domain of GGA1 could interact with the glutamates at position -3 and -2 ($E^{58,59}$) of the CD-MPR. The absence of positional restriction and irrelevance of the type of negative charge, which we suggested to range between -4 or -3 to -1, might explain the less ordered nature of the interaction upstream of the aspartate 0, which was found in the structural analysis. However, it should be considered, that only short peptides were used for the structural analysis, not enabling the formation of a three-dimensional structure, which might result in a different kind of interaction of residues upstream of the aspartate 0.

The kinetic analysis of the interaction between AP-1 and CD-MPR revealed the involvement of the four glutamates, $E^{55,56,58,59}$ of the CK2 site of the CD-MPR. Mutation of each pair of glutamates to alanines led to a 3-fold increase of the rate-constant of AP-1 binding. Comparable to the GGA1 binding, phosphorylation of the serine only had a negligible effect on CD-MPR binding to AP-1. Thus, AP-1 and GGA1 had overlapping binding sites, both comprising the $E^{58,59}$ and both independent of the S^{57} phosphorylation. Interestingly the binding affinity of wild-type CD-MPR peptide to AP-1 was 2.4-fold lower than the affinity to GGA1. The overlapping binding sites combined with the different binding affinities for CD-MPR led to the conclusion that AP-1 is unable to bind to CD-MPR in the presence of GGA1, unless GGA1 is inactivated by phosphorylation. Thus, our results confirm the suggested model, where GGA1 relays CD-MPR to AP-1, with slight modifications concerning the regulation (Ghosh and Kornfeld, 2003b). Our model suggests that GGA1 binds to CD-MPR with high affinity, thereby blocking the overlapping binding sites and disabling AP-1 binding. Subsequent phosphorylation of GGA1 by CK2, which is associated with AP-1, causes autoinhibition of GGA1, followed by the release of CD-MPR, rendering the overlapping binding site accessible for the weaker binding to AP-1. The localization of both adaptors, GGA1 and AP-1, to the TGN supports our model (Klumperman et al., 1993; Boman et al., 2000; Dell'Angelica et al., 2000; Hirst et al., 2000). Thus, we suggest that both adaptors are involved in sorting in the TGN independent of cargo phosphorylation, regulated by different K_{DS} for the overlapping sorting signals.

6. Acknowledgement

We thank Dr. S. Kornfeld for generously providing the myc-GGA1pFB1 construct. The hybridoma cell line producing mAb 22D4 against CD-MPR was a kind gift of Dr. D. Messner. P. Nair and B. Schaub are acknowledged for critical reading of the manuscript. We acknowledge Dr. E. Berger for continuous support and critical reading the manuscript. This work was supported in part by a Prof. Dr. Max Cloetta Fellowship to Jack Rohrer and Swiss National Science Foundation Grant 31-67274.

General Discussion

The CD-MPR cycles between the TGN, endosomes and the plasma membrane. In our study we tried to shed light on the characterization of two specific features of the CD-MPR - the palmitoylation and the phosphorylation of its cytoplasmic tail.

1. Results

1.1 Why are we interested in the Palmitoylation and Phosphorylation?

Once a membrane protein acquires its three-dimensional structure with its oligosaccharides correctly processed and leaves the TGN, some options remain to reversibly change the characteristics of the protein. Such reversible modifications of proteins present a potential regulatory mechanism. There are several types of modifications that can alter the feature of proteins, which include oligomerization, conformational changes upon protein-protein interaction or a change in pH and reversible post-translational modifications. Both palmitoylation and phosphorylation are reversible post-translational modifications, which are known to be used to regulate functions of proteins. Since the CD-MPR undergoes both palmitoylation and phosphorylation, it is of interest to analyze whether they perform a regulatory function in the CD-MPR.

1.2 Our hypothesis: Palmitoylation as a regulator of the sorting signals

The trafficking itinerary of the CD-MPR begins at the TGN, where the receptor binds the lysosomal enzymes through the M6P tags and continues through early endosomes and late endosomes. In late endosomes the CD-MPR dissociates from the lysosomal enzyme. The lysosomal hydrolases get transferred to the lysosomes, probably by partial fusions between late endosomes and lysosomes. In contrast, the CD-MPR must avoid delivery to lysosomes, since missorting to lysosomes would result in its rapid degradation. Being only one transport step away from degradation, one would expect that the CD-MPR has a rapid turnover. Instead, the half-life is more than 40 h, indicating that the receptor is efficiently transported out of late endosomes. Two motifs were found to be crucial to avoid delivery to lysosomes – the diaromatic motif $F^{18}W^{19}$ and the palmitoylated C^{34} (Schweizer et al., 1996; Schweizer et al., 1997). Avoidance of lysosomal degradation depends on the presence of both the palmitoylation and the diaromatic signal. The membrane anchoring of the palmitoylation presumably influences the diaromatic motif, which is probably better exposed to interacting proteins in the palmitoylated form comparable to the non-palmitoylated form of the CD-MPR. Our hypothesis suggests that the reversible palmitoylation

General Discussion

presents a regulatory mechanism for sorting signals in the cytoplasmic tail of the CD-MPR. This hypothesis would require that, palmitoylation occurs enzymatically and secondly, that palmitoylation takes place in an organelle that precedes the late endosomes in the trafficking itinerary of the CD-MPR. Our findings of a palmitoyltransferase that palmitoylates the CD-MPR and which cycles between endosomes and the plasma membrane fulfills both requirements (see Part I). Furthermore, the localization of the palmitoyltransferase is optimal to ensure the presence of the palmitoylated C^{34} in late endosomes. Altogether, our findings on the palmitoyltransferase support our hypothesis.

1.3 The cytoplasmic tail of the CD-MPR for its sorting in the TGN

The reports about the involvement of the CK2 site of the CD-MPR in the sorting of cathepsin D are controversial. The suggested functions of the phosphorylation of S^{57} included surface delivery, possibly through a sorting step in the endosomes and binding to AP-1 (see also chapter: *The phosphorylation of Ser57 and the CK2 site*, page 41) (Mauxion et al., 1996; Breuer et al., 1997; Ghosh and Kornfeld, 2003a).

We wanted to analyze whether the phosphorylation and the CK2 site influences the interaction of the CD-MPR with GGA1 and with AP-1. We could show that for both GGA1 and AP-1 the binding to the CD-MPR is independent of phosphorylation of the S^{57} (see Part II). However, the glutamates surrounding the serine (E^{55}, E^{56}, E^{58}, E^{59}) play a crucial role in the interaction of the CD-MPR with GGA1 and AP-1. The GGA1 requires the E^{58} and E^{59} in addition to the known D^{61}-X-X-L^{64}-L^{65}, whereas the AP-1 requires all glutamates (E^{55}, E^{56}, E^{58}, E^{59}) for interaction with the CD-MPR. This results in partially overlapping binding sites in the CD-MPR for GGA1 and AP-1. Interestingly the binding affinity of the interaction between GGA1 and the CD-MPR is significantly higher than that of AP-1 with the CD-MPR. Therefore we suggest a modification of Kornfeld's model for sorting in the TGN, where GGA1 interacts with the CD-MPR and recruits it to clathrin-coated pits, AP-1 is recruited to the coated pits by interaction with GGA1. The AP-1-associated CK2 phosphorylates GGA1, causing the autoinhibition of GGA1 and the release of the CD-MPR. AP-1 then interacts with the glutamates of the CD-MPR and mediates the formation of the clathrin-coated vesicle. Thus, both GGA1 and AP-1 are suggested to be involved in the transport of the CD-MPR from the TGN to endosomes and their interaction with the CD-MPR is regulated by different affinities towards the partially overlapping binding site. Altogether these results indicate that the acidic cluster of the CK2 site, but not the phosphorylation of the S^{57} is required for the sorting in the TGN.

General Discussion

2. Problems encountered

2.1 Cloning of the palmitoyltransferase

Our approaches to purify the palmitoyltransferase were not successful. Although we managed to obtain a soluble form of activity by sonicating membranes in the absence of protease inhibitors, we failed to enrich the activity by subjecting the soluble activity to various chromatography purification steps.

A second approach evolved after the publication of two potential yeast palmitoyltransferases, the Erf2p/Erf4p complex and Akr1p (Lobo et al., 2002; Roth et al., 2002). The sequence comparison revealed that both, Erf2p (the catalytic subunit of the complex) and Akr1p, contain a conserved domain, which includes the active site – the DHHC-cysteine-rich domain (CRD). It is part of a zinc-finger domain which is also referred to as the NEW1 domain. Through database searches we obtained twelve human homologues with the conserved NEW1 domain (see Figure 20). Of these twelve homologues, eight proteins contain a putative internalization motif, indicating that these proteins might be internalized from the plasma membrane to the endosomes and therefore might cycle between the plasma membrane and endosomes. Given that the palmitoyltransferase of the CD-MPR cycles between the plasma membrane and endosomes, these eight NEW1 homologues are candidates that ought to be tested for palmitoyltransferase activity on the CD-MPR.

Upon the first report about Erf2p, we cloned the human homologue hErf2, expressed it in HeLa cells and tested the extract of these cells in the *in vitro* palmitoylation assay. Unfortunately no enhanced palmitoylation of the CD-MPR was observed. This negative result was not entirely unexpected as in yeast the palmitoyltransferase activity of Erf2p depends on Erf4p for which we could not find a human homologue to co-express in the cells. In any case, hErf2 is not the most promising candidate for palmitoylating the CD-MPR because in yeast Erf2p is supposed to be localized in the ER and the potential internalization motif of hErf2 is only weak (dileucine motif).

Our goal for the future is to clone all eight NEW1 homologues, containing a putative internalization signal (see Figure 20) and test them for palmitoyltransferase activity on the CD-MPR.

3. Open questions

3.1 The function of the non-palmitoylated form of the CD-MPR

One open question is: if the CD-MPR is reversibly palmitoylated to regulate the sorting signals in its cytoplasmic tail, is there a sorting step that favors the non-palmitoylated form of the CD-MPR? The answer is: we do not know.

```
                                                              TMD   Internalization  Name
                1↓↓        ↓↓↓  ↓     ↓                       50    Motif
gi29791459  rycskcqlikpdra   csacdscilkmd  cpwv  cv fs yk ffllfllyal  FxxxxF,Yxx♦
gi28202111  rycdrcqlikpdrc   csvcdkcilkmd  cpwv  cv fs yk ffllflaysl  FxxxxF,Yxx♦
gi21450653  rfcdrc likpdrc   csvcamcvlkmd  cpwv  ci fs yk fflqflaysv  FxxxxF
gi28202103  kwcatcrfyrppros  csvcd cveafdh cpwv  ci rr yr ffflfll.sl  4x Yxx♦, LL
gi32698692  kwcatc fyrpprcs  csvcd cvedfd  cpwv  ci rr yr fflfll.sl   2x Yxx♦
gi24371241  kycftckifrppras  cslcd cverfdh cpwvg cv kr yr ffymfil.sl  2x Yxx♦
gi24371272  kycftckifrppras  cslcd cverfdh cpwvg cv kr yr ffymfil.sl  2x LL
gi28202113  kycytckifrppras  csicd cverfdh cpwvg cv kr yr ffylfil.sl  LL               hErf2
gi22041784  kycftckmfrpprts  csvcd cverfdh cpwvg cv rr yr ffyafil.sl  -
gi29244581  ifcstclirkpvrsk  cqvc rciakfdh cpwvg cv ag  yr fm yylffll 3x Yxx♦          hAkr1,Hip14
gi10834672  sickkciypkpart   csic rcvlkmd  cpwl  cv ny  r yffsfcffmt  -
gi14165541  sickkciypkpart   csic rcvlkmd  cpwl  cv y   r yffsfcffmt  YxYxx♦

consensus   kyc-tc-♦♦kp-r--  cs♦cd-cv-+fd  cpwv- cv --- yr yf♦-f♦l-sl
```

Figure 20: Sequence alignment of human NEW1-domain homologues. The amino acids are depicted in single code letters and the consensus sequence is shown in the two bottom lines. The alignment is restricted to the NEW1 domain. The conserved dhhc sequence is marked by the yellow box and arrow, the conserved cysteines are marked with a purple arrow. The TMD is marked with a green box. The internalization sequences found in the complete sequence and the names, if existent, are indicated next to the sequence. The promising candidates with good internalization sequences are highlighted in green.

The CD-MPR itinerary includes the TGN, the plasma membrane and endosomes. In late endosomes palmitoylation is required, but what about the plasma membrane and the TGN? We showed that the mutant CD-MPR, which is almost exclusively localized at the plasma membrane, is palmitoylated, indicating that the palmitoyltransferase is active at the plasma membrane and might also palmitoylate the portion of wild-type CD-MPR, which at steady-state resides at the plasma membrane (see Part I). Thus, a possible preference of the non-palmitoylated form for efficient internalization would be unreasonable. However, the main internalization motif of the CD-MPR (F^{13}-X-X-X-X-F^{18}) overlaps with the diaromatic motif ($F^{18}W^{19}$) and according to our hypothesis, the exposure of the internalization motif, like the diaromatic motif should be affected by the conformational change upon palmitoylation. Thus, one would expect that the palmitoylation might also be required for internalization. But the requirements for the structural environment of a sorting signal might also vary depending on the interacting protein(s). To address this question, we should compare the internalization rates of the wild-type CD-MPR and the palmitoylation-deficient CD-MPR-$C^{30,34}$A.

In the TGN, the CD-MPR requires GGA and most likely AP-1 for efficient transport to endosomes. We could show that the interaction between the CD-MPR and GGA1 is independent of palmitoylation, since GGA1 bound equally well to the wild-type CD-MPR and $C^{30,34}$A (see Part II). Furthermore we demonstrated that AP-1 requires the acidic cluster around the CK2 phosphorylation site (E^{55}, E^{56}, E^{58}, E^{59}) for interaction (see Part II). However, the analysis was done using a peptide, which does not include the cysteines. Höning and colleagues published that AP-1 interacts with two distinct sites in the CD-MPR, one comprising the amino acids 49 to 67, which is the same region we tested together with Stefan Höning (see Part II). The other binding site for AP-1 spans the amino acids 27 to 43 including the cysteines (C^{30}, C^{34}) that are palmitoylated *in vivo* and the basic residues

General Discussion

from 35 to 39 (R^{35}, K^{37}, R^{39}) that contribute to efficient palmitoylation (Schweizer et al., 1996; Höning et al., 1997). In the binding assay, the peptide was not palmitoylated, indicating that there might be a preference of AP-1 for the non-palmitoylated region between residues 27 to 34. This hypothesis should be tested by comparing the binding affinity of AP-1 to palmitoylated and non-palmitoylated peptides. Another question would be: if there are two AP-1 binding sites *in vitro*, are both equally involved in AP-1 binding *in vivo*?

Concluding the discussion about the function of the non-palmitoylated form of CD-MPR, it would be interesting to test the internalization rate of the CD-MPR-$C^{30,34}$A and to analyze the binding affinity of AP-1 to palmitoylated versus non-palmitoylated forms of CD-MPR peptides, or better yet, to the full-length receptor.

3.2 Palmitoylation and the diaromatic motif: TIP47?

The palmitoylation of C^{34} and the diaromatic motif $F^{18}W^{19}$ are both required for efficient transport from endosomes to the TGN (Schweizer et al., 1996; Schweizer et al., 1997). TIP47 was found to interact specifically with the diaromatic motif of the CD-MPR and was required for the transport from late endosomes to the TGN (see also chapter: *TIP47*, page 35) (Diaz and Pfeffer, 1998). Interestingly, the bait that was used to identify TIP47 in a yeast two-hybrid screen comprised the cytoplasmic tail of the CD-MPR fused to the GAL4 DNA binding domain. Thus, the CD-MPR did not contain the TMD and the palmitoylation state of this soluble fusion construct was not known. It is surprising that TIP47 bound to this fusion protein, considering that the structural requirements for the functional diaromatic motif are quite strict: the palmitoylation of C^{34} and the specific distance of $F^{18}W^{19}$ from the TMD. Mutant CD-MPRs with insertion or deletion of 5 amino acids between the diaromatic motif and the TMD resulted in missorting of the receptor to the lysosomes (Schweizer et al., 1997). Whether the binding of TIP47 to the CD-MPR depends on the same structural requirements of the diaromatic motif is not known. It would be interesting to test the binding of TIP47 to the following mutant CD-MPRs in an *in vitro* assay: CD-MPR-$C^{30,34}$A and CD-MPR with deletion or insertion of 5 amino acids between TMD and $F^{18}W^{19}$ (Nair et al., 2003). This experiment would tell us whether TIP47 is the only enzyme responsible for avoidance of lysosomal degradation of the receptor, or whether there might be additional proteins involved, that would together account for the requirements of the CD-MPR to be efficiently sorted out of the late endosomes, these being the palmitoylation and the diaromatic motif at the correct distance from the TMD.

3.3 Other candidate adaptor proteins for the transport from endosomes to the TGN?

Data suggesting that other proteins are involved in the transport of the CD-MPR from endosomes to the TGN exists in the literature. In mouse fibroblasts, deficient in AP-1, the CD-MPR is redistributed to early endosomes and fails to recycle back from endosomes to the TGN (Meyer et al., 2000). Furthermore, the *in vitro* transport step from endosomes to the TGN is dependent on membrane-bound AP-1 and on cytosolic AP-3, but not on TIP47 in mouse fibroblasts (Medigeshi and Schu, 2003). Multiple binding sites for the μ3 subunit of AP-3 were identified in the CD-MPR *in vitro*, including the tyrosine motif (Y^{45}-X-X-V^{48}), the acidic cluster of the CK2 site (E^{55}, E^{56}, E^{58}, E^{59}) and the dileucine motif (L^{64}-L^{65}) (Storch and Braulke, 2001). These results indicate that AP-1 and AP-3 are required for the retrograde transport of the receptor from endosomes to the TGN.

In addition to TIP47 more proteins might be involved in the transport of the CD-MPR from endosomes to the TGN, such as AP-1 and AP-3. One possibility is that different proteins or adaptors mediate the transport originating on distinct sites within the endosomal system. It is suggested that the CD-MPR is transported back to the TGN from early endosomes, as well as from late endosomes, possibly requiring a different subset of proteins for the two transport steps. This would also explain the controversial reports about the requirement for TIP47 in the transport of the CD-MPR from endosomes to the TGN in two different cell lines (Diaz and Pfeffer, 1998; Medigeshi and Schu, 2003). The steady-state distribution of the CD-MPR varies among cell lines and it is possible that one cell line predominantly uses the EE to TGN transport, whereas the other cell line prefers the LE to TGN transport.

Another candidate for the transport from endosomes to the TGN of the CD-MPR is PACS-1 (see also chapter: *PACS-1*, page 34). It binds to acidic clusters of CK2 sites in cargo and is required for the transport of furin and CI-MPR to the TGN (Wan et al., 1998). The interaction of PACS-1 with furin is dependent on the phosphorylation by CK2, which is not the case with the CI-MPR. PACS-1 requires AP-1 and/or AP-3 for the transport of cargo from endosomes to the TGN (Crump et al., 2001). Whether PACS-1 interacts with the CD-MPR is not known. The dependence of PACS-1 on AP-1 and/or AP-3 would match the requirement of the CD-MPR for AP-1 and AP-3 in the transport from endosomes to the TGN (see above). Thus, it might be interesting to analyze whether the CD-MPR and PACS-1 interact with each other and if this is indeed the case, whether the interaction is dependent on phosphorylation of the CK2 site.

3.4 Where does the missorting of the CD-MPR occur?

There are two ways how a membrane protein can end up in the lysosome: on the limiting membrane of the lysosome or on internal vesicles that derive from MVB/LE and are transferred to

General Discussion

the lysosome either by partial or by complete fusion (Storrie and Desjardins, 1996; Luzio et al., 2000). Is the missorted CD-MPR (lacking the cysteines or the diaromatic motif) localized on internal membranes or in the outer membrane of the lysosome? Co-immunofluorescence of missorted CD-MPR with LAMP-1, which is on the outer membrane, revealed that both colocalize in lysosomal, doughnut-shaped structures, however they partially segregate into subdomains on lysosomes (Nair et al., 2003). It was suggested that lysosomes contain subdomains, which are involved in degradation and other subdomains for export, as it was shown for LAMP-1 (Lippincott-Schwartz and Fambrough, 1987; Furuno et al., 1989; Akasaki et al., 1993). However, these different subdomains might also represent the internal vesicles and the outer membrane. Thus, missorting of the receptor could already start in early endosomes. This implies that the palmitoylation and the diaromatic motif could also be involved in the retention of the CD-MPR to the outer membrane on EE and in the avoidance of the invagination of the receptor into intralumenal vesicles. Hence, it might be an idea to test this alternative missorting to lysosomes and look for interaction partners of the CD-MPR in EE, which are involved in the sorting of cargo.

3.5 CD-MPR at the plasma membrane, why?

10-20% of the CD-MPR is localized to the cell surface. The function of the CD-MPR at the plasma membrane is unclear or rather inexplicable, since the CD-MPR is not able to bind ligands at the cell surface. Furthermore, it is not known how the CD-MPR is delivered to the plasma membrane. It should not get there directly from the TGN or else the receptor would lose the ligand and thereby missort lysosomal enzymes. Thus, the CD-MPR is presumably transported to the plasma membrane from endosomes. However, it is not known whether every receptor molecule travels to the plasma membrane through a directed transport step from endosomes, or whether only part of the CD-MPRs are missorted to the plasma membrane. The major question remains, why is the CD-MPR at the plasma membrane at all? Is it misrouting or does it happen on purpose?

3.6 Effect of palmitoylation on phosphorylation and vice versa

The three-dimensional structure of the cytoplasmic tail of the CD-MPR is not known. Hence we do not know the effect of the membrane anchoring through the palmitate. In addition to the probable exposure of the diaromatic motif, there might be other signals that are affected by the putative conformational change. One interesting aspect is the influence of the palmitoylation on phosphorylation and vice versa. Preliminary data indicated that the lack of palmitoylation induced a reduction in phosphorylation, but palmitoylation seems to be unaffected by phosphorylation. This result has to be reconfirmed and the effect has to be further investigated. These data suggest that the palmitoylation might not only regulate the diaromatic motif, but also other signals within the cytoplasmic tail. This raises the question again, which adaptor protein accounts for the requirements of palmitoylation. In addition to proteins that interact with the diaromatic motif, such as TIP47, there might be proteins involved in binding to the phosphorylated serine. This hypothesis again demands for the analysis of the interaction between PACS-1 and the CD-MPR.

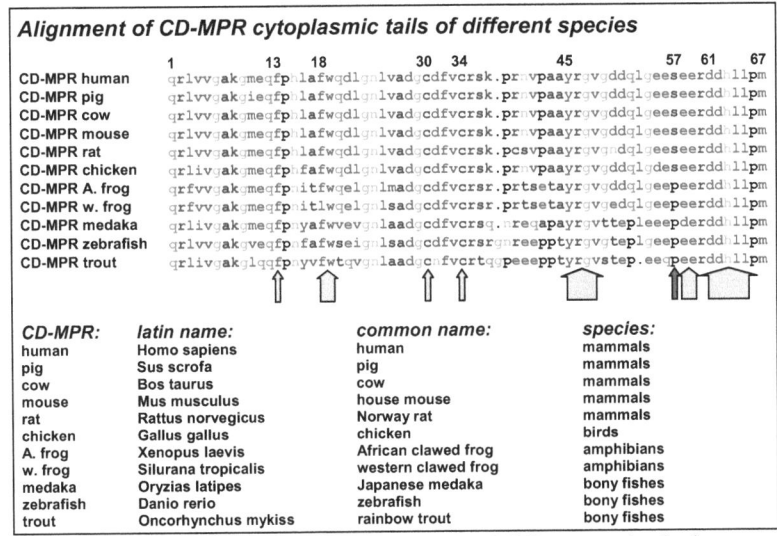

Figure 21: Alignment of CD-MPR cytoplasmic tails of different species. In the upper part, amino acid sequence is depicted in single letter code. Signals, which are important for trafficking, are marked with a yellow arrow. The proline or serine at position 57 is marked with the red arrow. Common names, latin names and the species of the sequences are shown in the lower part.

3.7 What about the phosphorylation of the S^{57}?

Phosphorylation is not involved in sorting in the TGN, but whether phosphorylation is required at the plasma membrane or in the endosomes is not known. If indeed the palmitoylation has an effect on phosphorylation, then phosphorylation might also be required for the transport of the CD-MPR from endosomes to the TGN. Thus, future experiments would include the confirmation of the effect of palmitoylation on phosphorylation and an analysis whether the phosphorylation mutants are missorted to lysosomes.

Missorting to lysosomes of mutant CD-MPRs and subsequent degradation would lead to a drastic reduction of their steady-state levels, and thus would cause missorting of cathepsin D, as it has been shown for the CD-MPR-$C^{30,34}$A (Schweizer et al., 1996). Although studies about the CK2 phosphorylation sites revealed controversial results on the sorting of cathepsin D concerning the glutamates, none of the investigators observed missorting of cathepsin D when only the serine was replaced. This indicates that the serine might not be essential for sorting of the receptor (Johnson and Kornfeld, 1992b; Mauxion et al., 1996; Breuer et al., 1997).

General Discussion

A striking fact emerges from the sequence comparison of the cytoplasmic tail of the CD-MPR between different species (see Figure 21). The cytoplasmic tail of the CD-MPR is extremely conserved among species, showing even 100% identity between human, mouse, pig and cow. When comparing the CK2 sites between the species, it is interesting that all bony fishes and amphibians lack the serine and instead, have a proline at position 57. This fact raises doubts about the functional relevance of the phosphorylation of the serine.

4. *Concluding remarks*

The trafficking and the function of the CD-MPR have been extensively investigated. The receptor cycles between TGN, plasma membrane and endosomes. Various sorting signals are identified in the cytoplasmic tail that mediate the distinct sorting steps.

We contributed to the understanding of certain trafficking steps and sorting signals in the CD-MPR, including the characterization of the palmitoylation and the influence of the phosphorylation and the acidic cluster of the CK2 site in sorting in the TGN.

However, many open questions remain: What is the function of the non-palmitoylated CD-MPR? Does the palmitoylation have an influence on more than one signal of the cytoplasmic tail? Is TIP47 the only protein that prevents missorting of the CD-MPR? Alternatively, are many enzymes involved? Are there different routes from endosomes back to the TGN? What is the function of phosphorylation? Is there a function for phosphorylation at all? Why is part the CD-MPR at the cell surface? How and from where is it delivered to the plasma membrane?

The established methods and all the mutant CD-MPR constructs that we made would allow us to address some of these questions using some of the described approaches. Therefore we could get closer to the ultimate goal – the complete understanding of the trafficking of the CD-MPR.

References

Adam, M., A. Rodriguez, C. Turbide, J. Larrick, E. Meighen, and R.M. Johnstone. 1984. In vitro acylation of the transferrin receptor. *J Biol Chem.* 259:15460-3.

Aguilar, R.C., M. Boehm, I. Gorshkova, R.J. Crouch, K. Tomita, T. Saito, H. Ohno, and J.S. Bonifacino. 2001. Signal-binding specificity of the mu4 subunit of the adaptor protein complex AP-4. *J Biol Chem.* 276:13145-52.

Ahle, S., A. Mann, U. Eichelsbacher, and E. Ungewickell. 1988. Structural relationships between clathrin assembly proteins from the Golgi and the plasma membrane. *Embo J.* 7:919-29.

Akasaki, K., M. Fukuzawa, H. Kinoshita, K. Furuno, and H. Tsuji. 1993. Cycling of two endogenous lysosomal membrane proteins, lamp-2 and acid phosphatase, between the cell surface and lysosomes in cultured rat hepatocytes. *J Biochem (Tokyo).* 114:598-604.

Alvarez, E., N. Girones, and R.J. Davis. 1990. Inhibition of the receptor-mediated endocytosis of diferric transferrin is associated with the covalent modification of the transferrin receptor with palmitic acid. *J Biol Chem.* 265:16644-55.

Amor, J.C., D.H. Harrison, R.A. Kahn, and D. Ringe. 1994. Structure of the human ADP-ribosylation factor 1 complexed with GDP. *Nature.* 372:704-8.

Aniento, F., N. Emans, G. Griffiths, and J. Gruenberg. 1993. Cytoplasmic dynein-dependent vesicular transport from early to late endosomes [published erratum appears in J Cell Biol 1994 Feb;124(3):397]. *J Cell Biol.* 123:1373-87.

Arvan, P., and D. Castle. 1998. Sorting and storage during secretory granule biogenesis: looking backward and looking forward. *Biochem J.* 332 (Pt 3):593-610.

Babst, M., D.J. Katzmann, E.J. Estepa-Sabal, T. Meerloo, and S.D. Emr. 2002. Escrt-III: an endosome-associated heterooligomeric protein complex required for mvb sorting. *Dev Cell.* 3:271-82.

Bache, K.G., A. Brech, A. Mehlum, and H. Stenmark. 2003a. Hrs regulates multivesicular body formation via ESCRT recruitment to endosomes. *J Cell Biol.* 162:435-42.

Bache, K.G., C. Raiborg, A. Mehlum, and H. Stenmark. 2003b. STAM and Hrs are subunits of a multivalent ubiquitin-binding complex on early endosomes. *J Biol Chem.* 278:12513-21.

Bano, M.C., C.S. Jackson, and A.I. Magee. 1998. Pseudo-enzymatic S-acylation of a myristoylated yes protein tyrosine kinase peptide in vitro may reflect non-enzymatic S-acylation in vivo. *Biochem J.* 330 (Pt 2):723-31.

Bao, M., J.L. Booth, B.J. Elmendorf, and W.M. Canfield. 1996a. Bovine UDP-N-acetylglucosamine:lysosomal-enzyme N-acetylglucosamine-1-phosphotransferase. I. Purification and subunit structure. *J Biol Chem.* 271:31437-45.

Bao, M., B.J. Elmendorf, J.L. Booth, R.R. Drake, and W.M. Canfield. 1996b. Bovine UDP-N-acetylglucosamine:lysosomal-enzyme N-acetylglucosamine-1-phosphotransferase. II. Enzymatic characterization and identification of the catalytic subunit. *J Biol Chem.* 271:31446-51.

Barlowe, C., L. Orci, T. Yeung, M. Hosobuchi, S. Hamamoto, N. Salama, M.F. Rexach, M. Ravazzola, M. Amherdt, and R. Schekman. 1994. COPII: a membrane coat formed by Sec proteins that drive vesicle budding from the endoplasmic reticulum. *Cell.* 77:895-907.

Barr, V.A., S.A. Phillips, S.I. Taylor, and C.R. Haft. 2000. Overexpression of a novel sorting nexin, SNX15, affects endosome morphology and protein trafficking. *Traffic.* 1:904-916.

Bean, A.J., S. Davanger, M.F. Chou, B. Gerhardt, S. Tsujimoto, and Y.C. Chang. 2000. Hrs-2 regulates receptor-mediated endocytosis via interactions with Eps15. *J Biol Chem.* 275:15271-15278.

Belanger, C., H. Ansanay, R. Qanbar, and M. Bouvier. 2001. Primary sequence requirements for S-acylation of beta(2)-adrenergic receptor peptides. *FEBS Lett.* 499:59-64.

Benmerah, A., B. Begue, A. Dautry-Varsat, and N. Cerf-Bensussan. 1996. The ear of alpha-adaptin interacts with the COOH-terminal domain of the Eps 15 protein. *J Biol Chem.* 271:12111-6.

Benmerah, A., C. Lamaze, B. Begue, S.L. Schmid, A. Dautryvarsat, and N. Cerfbensussan. 1998. AP-2/Eps15 interaction is required for receptor-mediated endocytosis. *J Cell Biol.* 140:1055-1062.

Berger, E.G., E. Aegerter, T. Mandel, and H.P. Hauri. 1986. Monoclonal antibodies to soluble, human milk galactosyltransferase (lactose synthase A protein). *Carbohydr Res.* 149:23-33.

Blanchard, F., L. Duplomb, S. Raher, P. Vusio, B. Hoflack, Y. Jacques, and A. Godard. 1999. Mannose 6-Phosphate/Insulin-like growth factor II receptor mediates internalization and degradation of leukemia inhibitory factor but not signal transduction. *J Biol Chem.* 274:24685-93.

Blanpain, C., V. Wittamer, J.M. Vanderwinden, A. Boom, B. Renneboog, B. Lee, E. Le Poul, L. El Asmar, C. Govaerts, G. Vassart, R.W. Doms, and M. Parmentier. 2001. Palmitoylation of CCR5 is critical for receptor trafficking and efficient activation of intracellular signaling pathways. *J Biol Chem.* 276:23795-804.

Bock, J.B., H.T. Matern, A.A. Peden, and R.H. Scheller. 2001. A genomic perspective on membrane compartment organization. *Nature.* 409:839-41.

Boll, W., H. Ohno, Z. Songyang, I. Rapoport, L.C. Cantley, J.S. Bonifacino, and T. Kirchhausen. 1996. Sequence requirements for the recognition of tyrosine-based endocytic signals by clathrin AP-2 complexes. *Embo J.* 15:5789-95.

Boll, W., I. Rapoport, C. Brunner, Y. Modis, S. Prehn, and T. Kirchhausen. 2002. The mu2 subunit of the clathrin adaptor AP-2 binds to FDNPVY and YppO sorting signals at distinct sites. *Traffic.* 3:590-600.

Boman, A.L., C. Zhang, X. Zhu, and R.A. Kahn. 2000. A family of ADP-ribosylation factor effectors that can alter membrane transport through the trans-Golgi. *Mol Biol Cell.* 11:1241-55.

Bonatti, S., G. Migliaccio, and K. Simons. 1989. Palmitoylation of viral membrane glycoproteins takes place after exit from the endoplasmic reticulum. *J Biol Chem.* 264:12590-5.

Bonifacino, J.S., and J. Lippincott-Schwartz. 2003. Coat proteins: shaping membrane transport. *Nat Rev Mol Cell Biol.* 4:409-14.

Braulke, T., and G. Mieskes. 1992. Role of protein phosphatases in insulin-like growth factor II (IGF II)-stimulated mannose 6-phosphate/IGF II receptor redistribution. *J Biol Chem.* 267:17347-53.

Bremnes, T., V. Lauvrak, B. Lindqvist, and O. Bakke. 1998. A region from the medium chain adaptor subunit (mu) recognizes leucine- and tyrosine-based sorting signals. *J Biol Chem.* 273:8638-8645.

Bresciani, R., K. Denzer, R. Pohlmann, and K. von Figura. 1997. The 46 kDa mannose-6-phosphate receptor contains a signal for basolateral sorting within the 19 juxtamembrane cytosolic residues. *Biochem J.* 327:811-8.

Brett, T.J., L.M. Traub, and D.H. Fremont. 2002. Accessory protein recruitment motifs in clathrin-mediated endocytosis. *Structure (Camb).* 10:797-809.

Breuer, P., C. Körner, C. Boker, A. Herzog, R. Pohlmann, and T. Braulke. 1997. Serine phosphorylation site of the 46-kDa mannose 6- phosphate receptor is required for transport to the plasma membrane in Madin-Darby canine kidney and mouse fibroblast cells. *Mol Biol Cell.* 8:567-576.

Brown, W.J., D.B. DeWald, S.D. Emr, H. Plutner, and W.E. Balch. 1995. Role for phosphatidylinositol 3-kinase in the sorting and transport of newly synthesized lysosomal enzymes in mammalian cells. *J Cell Biol.* 130:781-96.

Burgoyne, R.D., and A. Morgan. 2003. Secretory granule exocytosis. *Physiol Rev.* 83:581-632.

Calero, G., P. Gupta, M.C. Nonato, S. Tandel, E.R. Biehl, S.L. Hofmann, and J. Clardy. 2003. The crystal structure of palmitoyl protein thioesterase-2 (PPT2) reveals the basis for divergent substrate specificities of the two lysosomal thioesterases, PPT1 and PPT2. *J Biol Chem.* 278:37957-64.

References

Camp, L.A., and S.L. Hofmann. 1993. Purification and properties of a palmitoyl-protein thioesterase that cleaves palmitate from H-Ras. *J Biol Chem.* 268:22566-74.

Camp, L.A., L.A. Verkruysc, S.J. Afendis, C.A. Slaughter, and S.L. Hofmann. 1994. Molecular cloning and expression of palmitoyl-protein thioesterase. *J Biol Chem.* 269:23212-9.

Canfield, W.M., K.F. Johnson, R.D. Ye, W. Gregory, and S. Kornfeld. 1991. Localization of the signal for rapid internalization of the bovine cation-independent mannose 6-phosphate/insulin-like growth factor-II receptor to amino acids 24-29 of the cytoplasmic tail. *J Biol Chem.* 266:5682-8.

Carroll, K.S., J. Hanna, I. Simon, J. Krise, P. Barbero, and S.R. Pfeffer. 2001. Role of Rab9 GTPase in facilitating receptor recruitment by TIP47. *Science.* 292:1373-6.

Chamoun, Z., R.K. Mann, D. Nellen, D.P. von Kessler, M. Bellotto, P.A. Beachy, and K. Basler. 2001. Skinny hedgehog, an acyltransferase required for palmitoylation and activity of the hedgehog signal. *Science.* 293:2080-4.

Chardin, P., S. Paris, B. Antonny, S. Robineau, S. Beraud-Dufour, C.L. Jackson, and M. Chabre. 1996. A human exchange factor for ARF contains Sec7- and pleckstrin-homology domains. *Nature.* 384:481-4.

Chavrier, P., R.G. Parton, H.P. Hauri, K. Simons, and M. Zerial. 1990. Localization of low molecular weight GTP binding proteins to exocytic and endocytic compartments. *Cell.* 62:317-29.

Chen, C.A., and D.R. Manning. 2001. Regulation of G proteins by covalent modification. *Oncogene.* 20:1643-52.

Chen, H., S. Fre, V.I. Slepnev, M.R. Capua, K. Takei, M.H. Butler, P.P. Difiore, and P. Decamilli. 1998. Epsin is an EH-domain-binding protein implicated in clathrin-mediated endocytosis. *Nature.* 394:793-797.

Chen, H.J., J. Yuan, and P. Lobel. 1997. Systematic mutational analysis of the cation-independent mannose 6-phosphate/insulin-like growth factor II receptor cytoplasmic domain - An acidic cluster containing a key aspartate is important for function in lysosomal enzyme sorting. *J Biol Chem.* 272:7003-7012.

Chidambaram, S., N. Mullers, K. Wiederhold, V. Haucke, and G. Fischer Von Mollard. 2003. Specific interaction between SNAREs and ENTH domains of epsin-related proteins in TGN to endosome transport. *J Biol Chem.*

Chin, L.S., M.C. Raynor, X. Wei, H.Q. Chen, and L. Li. 2001. Hrs interacts with sorting nexin 1 and regulates degradation of epidermal growth factor receptor. *J Biol Chem.* 276:7069-78.

Christoforidis, S., M. Miaczynska, K. Ashman, M. Wilm, L.Y. Zhao, S.C. Yip, M.D. Waterfield, J.M. Backer, and M. Zerial. 1999. Phosphatidylinositol-3-OH kinases are Rab5 effectors. *Nat Cell Biol.* 1:249-252.

Collins, B.M., A.J. McCoy, H.M. Kent, P.R. Evans, and D.J. Owen. 2002. Molecular architecture and functional model of the endocytic AP2 complex. *Cell.* 109:523-35.

Collins, B.M., P.J. Watson, and D.J. Owen. 2003. The structure of the GGA1-GAT domain reveals the molecular basis for ARF binding and membrane association of GGAs. *Dev Cell.* 4:321-32.

Cong, M., S.J. Perry, L.A. Hu, P.I. Hanson, A. Claing, and R.J. Lefkowitz. 2001. Binding of the beta2 adrenergic receptor to N-ethylmaleimide-sensitive factor regulates receptor recycling. *J Biol Chem.* 276:45145-52.

Corvi, M.M., C.L. Soltys, and L.G. Berthiaume. 2001. Regulation of mitochondrial carbamoyl-phosphate synthetase 1 activity by active site fatty acylation. *J Biol Chem.* 276:45704-12.

Cremona, O., G. Di Paolo, M.R. Wenk, A. Luthi, W.T. Kim, K. Takei, L. Daniell, Y. Nemoto, S.B. Shears, R.A. Flavell, D.A. McCormick, and P. De Camilli. 1999. Essential role of phosphoinositide metabolism in synaptic vesicle recycling. *Cell.* 99:179-88.

Crump, C.M., Y. Xiang, L. Thomas, F. Gu, C. Austin, S.A. Tooze, and G. Thomas. 2001. PACS-1 binding to adaptors is required for acidic cluster motif-mediated protein traffic. *Embo J.* 20:2191-201.

References

Cupers, P., A.P. Jadhav, and T. Kirchhausen. 1998. Assembly of clathrin coats disrupts the association between Eps15 and AP-2 adaptors. *J Biol Chem.* 273:1847-50.

Dahms, N.M., and M.K. Hancock. 2002. P-type lectins. *Biochim Biophys Acta.* 1572:317-40.

Dahms, N.M., P.A. Rose, J.D. Molkentin, Y. Zhang, and M.A. Brzycki. 1993. The bovine mannose 6-phosphate/insulin-like growth factor II receptor. The role of arginine residues in mannose 6-phosphate binding. *J Biol Chem.* 268:5457-63.

Das, A.K., B. Dasgupta, R. Bhattacharya, and J. Basu. 1997. Purification and biochemical characterization of a protein-palmitoyl acyltransferase from human erythrocytes. *J Biol Chem.* 272:11021-5.

David, C., P.S. McPherson, O. Mundigl, and P. de Camilli. 1996. A role of amphiphysin in synaptic vesicle endocytosis suggested by its binding to dynamin in nerve terminals. *Proc Natl Acad Sci U S A.* 93:331-5.

Davis, A.F., J. Bai, D. Fasshauer, M.J. Wolowick, J.L. Lewis, and E.R. Chapman. 1999. Kinetics of synaptotagmin responses to Ca2+ and assembly with the core SNARE complex onto membranes. *Neuron.* 24:363-76.

de Beer, T., R.E. Carter, K.E. Lobel-Rice, A. Sorkin, and M. Overduin. 1998. Structure and Asn-Pro-Phe binding pocket of the Eps15 homology domain. *Science.* 281:1357-60.

De Camilli, P., H. Chen, J. Hyman, E. Panepucci, A. Bateman, and A.T. Brunger. 2002. The ENTH domain. *FEBS Lett.* 513:11-8.

Dell'Angelica, E.C., J. Klumperman, W. Stoorvogel, and J.S. Bonifacino. 1998. Association of the AP-3 adaptor complex with clathrin. *Science.* 280:431-4.

Dell'Angelica, E.C., C. Mullins, and J.S. Bonifacino. 1999a. AP-4, a novel protein complex related to clathrin adaptors. *J Biol Chem.* 274:7278-85.

Dell'Angelica, E.C., H. Ohno, C.E. Ooi, E. Rabinovich, K.W. Roche, and J.S. Bonifacino. 1997. AP-3: an adaptor-like protein complex with ubiquitous expression. *Embo J.* 16:917-28.

Dell'Angelica, E.C., R. Puertollano, C. Mullins, R.C. Aguilar, J.D. Vargas, L.M. Hartnell, and J.S. Bonifacino. 2000. GGAs: a family of ADP ribosylation factor-binding proteins related to adaptors and associated with the Golgi complex. *J Cell Biol.* 149:81-94.

Dell'Angelica, E.C., V. Shotelersuk, R.C. Aguilar, W.A. Gahl, and J.S. Bonifacino. 1999b. Altered trafficking of lysosomal proteins in Hermansky-Pudlak syndrome due to mutations in the beta 3A subunit of the AP-3 adaptor. *Mol Cell.* 3:11-21.

Denzer, K., B. Weber, A. Hille-Rehfeld, K. von Figura, and R. Pohlmann. 1997. Identification of three internalization sequences in the cytoplasmic tail of the 46 kDa mannose 6-phosphate receptor. *Biochem J.* 326:497-505.

Desnoyers, L., J.S. Anant, and M.C. Seabra. 1996. Geranylgeranylation of Rab proteins. *Biochem Soc Trans.* 24:699-703.

Devedjiev, Y., Z. Dauter, S.R. Kuznetsov, T.L. Jones, and Z.S. Derewenda. 2000. Crystal structure of the human acyl protein thioesterase I from a single X-ray data set to 1.5 A. *Structure Fold Des.* 8:1137-46.

Devi, G.R., J.C. Byrd, D.H. Slentz, and R.G. MacDonald. 1998. An insulin-like growth factor II (IGF-II) affinity-enhancing domain localized within extracytoplasmic repeat 13 of the IGF-II/mannose 6-phosphate receptor. *Mol Endocrinol.* 12:1661-72.

Diaz, E., and S.R. Pfeffer. 1998. TIP47: A cargo selection device for mannose 6-phosphate receptor trafficking. *Cell.* 93:433-443.

Dirac-Svejstrup, A.B., T. Sumizawa, and S.R. Pfeffer. 1997. Identification of a GDI displacement factor that releases endosomal Rab GTPases from Rab-GDI. *Embo J.* 16:465-472.

Dittie, A.S., L. Thomas, G. Thomas, and S.A. Tooze. 1997. Interaction of furin in immature secretory granules from neuroendocrine cells with the AP-1 adaptor complex is modulated by casein kinase II phosphorylation. *Embo J.* 16:4859-70.

Do, H., W.S. Lee, P. Ghosh, T. Hollowell, W. Canfield, and S. Kornfeld. 2002. Human mannose 6-phosphate-uncovering enzyme is synthesized as a proenzyme that is activated by the endoprotease furin. *J Biol Chem.* 277:29737-44.

References

Doi, T., H. Sugimoto, I. Arimoto, Y. Hiroaki, and Y. Fujiyoshi. 1999. Interactions of endothelin receptor subtypes A and B with Gi, Go, and Gq in reconstituted phospholipid vesicles. *Biochemistry*. 38:3090-9.

Donaldson, J.G. 2000. Filling in the GAPs in the ADP-ribosylation factor story. *Proc Natl Acad Sci U S A*. 97:3792-4.

Doray, B., K. Bruns, P. Ghosh, and S.A. Kornfeld. 2002a. Autoinhibition of the ligand-binding site of GGA1/3 VHS domains by an internal acidic cluster-dileucine motif. *Proc Natl Acad Sci U S A*. 99:8072-7.

Doray, B., P. Ghosh, J. Griffith, H.J. Geuze, and S. Kornfeld. 2002b. Cooperation of GGAs and AP-1 in packaging MPRs at the trans-Golgi network. *Science*. 297:1700-3.

Doray, B., and S. Kornfeld. 2001. Gamma subunit of the AP-1 adaptor complex binds clathrin: implications for cooperative binding in coated vesicle assembly. *Mol Biol Cell*. 12:1925-35.

Duncan, J.A., and A.G. Gilman. 1996. Autoacylation of G protein alpha subunits. *J Biol Chem*. 271:23594-600.

Duncan, J.A., and A.G. Gilman. 1998. A cytoplasmic acyl-protein thioesterase that removes palmitate from G protein alpha subunits and p21(RAS). *J Biol Chem*. 273:15830-7.

Duncan, J.A., and A.G. Gilman. 2002. Characterization of Saccharomyces cerevisiae acyl-protein thioesterase 1, the enzyme responsible for G protein alpha subunit deacylation in vivo. *J Biol Chem*. 277:31740-52.

Dunphy, J.T., W.K. Greentree, C.L. Manahan, and M.E. Linder. 1996. G-protein palmitoyltransferase activity is enriched in plasma membranes. *J Biol Chem*. 271:7154-9.

Dunphy, J.T., H. Schroeder, R. Leventis, W.K. Greentree, J.K. Knudsen, J.R. Silvius, and M.E. Linder. 2000. Differential effects of acyl-CoA binding protein on enzymatic and non-enzymatic thioacylation of protein and peptide substrates. *Bba Mol Cell Biol Lipids*. 1485:185-198.

Endo, K., T. Takeshita, H. Kasai, Y. Sasaki, N. Tanaka, H. Asao, K. Kikuchi, M. Yamada, M. Chenb, J.J. O'Shea, and K. Sugamura. 2000. STAM2, a new member of the STAM family, binding to the Janus kinases. *FEBS Lett*. 477:55-61.

Engqvist-Goldstein, A.E., R.A. Warren, M.M. Kessels, J.H. Keen, J. Heuser, and D.G. Drubin. 2001. The actin-binding protein Hip1R associates with clathrin during early stages of endocytosis and promotes clathrin assembly in vitro. *J Cell Biol*. 154:1209-23.

Fang, S., J.P. Jensen, R.L. Ludwig, K.H. Vousden, and A.M. Weissman. 2000. Mdm2 is a RING finger-dependent ubiquitin protein ligase for itself and p53. *J Biol Chem*. 275:8945-51.

Farazi, T.A., G. Waksman, and J.I. Gordon. 2001. The biology and enzymology of protein N-myristoylation. *J Biol Chem*. 276:39501-4.

Fasshauer, D., R.B. Sutton, A.T. Brunger, and R. Jahn. 1998. Conserved structural features of the synaptic fusion complex: SNARE proteins reclassified as Q- and R-SNAREs. *Proc Natl Acad Sci U S A*. 95:15781-6.

Feng, L.J., A.B. Seymour, S. Jiang, A. To, A.A. Peden, E.K. Novak, L.J. Zhen, M.E. Rusiniak, E.M. Eicher, M.S. Robinson, M.B. Gorin, and R.T. Swank. 1999. The beta 3A subunit gene (Ap3b1) of the AP-3 adaptor complex is altered in the mouse hypopigmentation mutant pearl, a model for Hermansky-Pudlak syndrome and night blindness. *Hum Mol Genet*. 8:323-330.

Fischer von Mollard, G., and T.H. Stevens. 1999. The Saccharomyces cerevisiae v-SNARE Vti1p is required for multiple membrane transport pathways to the vacuole. *Mol Biol Cell*. 10:1719-32.

Fishburn, C.S., P. Herzmark, J. Morales, and H.R. Bourne. 1999. Gbetagamma and palmitate target newly synthesized Galphaz to the plasma membrane. *J Biol Chem*. 274:18793-800.

Folsch, H., H. Ohno, J.S. Bonifacino, and I. Mellman. 1999. A novel clathrin adaptor complex mediates basolateral targeting in polarized epithelial cells. *Cell*. 99:189-98.

Folsch, H., M. Pypaert, P. Schu, and I. Mellman. 2001. Distribution and function of AP-1 clathrin adaptor complexes in polarized epithelial cells. *J Cell Biol*. 152:595-606.

References

Ford, M.G., I.G. Mills, B.J. Peter, Y. Vallis, G.J. Praefcke, P.R. Evans, and H.T. McMahon. 2002. Curvature of clathrin-coated pits driven by epsin. *Nature.* 419:361-6.

Ford, M.G.J., B.M.F. Pearse, M.K. Higgins, Y. Vallis, D.J. Owen, A. Gibson, C.R. Hopkins, P.R. Evans, and H.T. McMahon. 2001. Simultaneous binding of PtdIns(4,5)P-2 and clathrin by AP180 in the nucleation of clathrin lattices on membranes. *Science.* 291:1051-1055.

Furuno, K., S. Yano, K. Akasaki, Y. Tanaka, Y. Yamaguchi, H. Tsuji, M. Himeno, and K. Kato. 1989. Biochemical analysis of the movement of a major lysosomal membrane glycoprotein in the endocytic membrane system. *J Biochem (Tokyo).* 106:717-22.

Futter, C.E., A. Pearse, L.J. Hewlett, and C.R. Hopkins. 1996. Multivesicular endosomes containing internalized EGF-EGF receptor complexes mature and then fuse directly with lysosomes. *J Cell Biol.* 132:1011-23.

Garmroudi, F., G. Devi, D.H. Slentz, B.S. Schaffer, and R.G. MacDonald. 1996. Truncated forms of the insulin-like growth factor II (IGF-II)/mannose 6-phosphate receptor encompassing the IGF-II binding site: characterization of a point mutation that abolishes IGF-II binding. *Mol Endocrinol.* 10:642-51.

Gaullier, J.M., A. Simonsen, A. D'Arrigo, B. Bremnes, H. Stenmark, and R. Aasland. 1998. FYVE fingers bind PtdIns(3)P. *Nature.* 394:432-3.

Gerdes, H.H., and M.M. Glombik. 1999. Signal-mediated sorting to the regulated pathway of protein secretion. *Anat Anz.* 181:447-53.

Ghosh, P., N.M. Dahms, and S. Kornfeld. 2003. Mannose 6-phosphate receptors: new twists in the tale. *Nat Rev Mol Cell Biol.* 4:202-12.

Ghosh, P., and S. Kornfeld. 2003a. AP-1 binding to sorting signals and release from clathrin-coated vesicles is regulated by phosphorylation. *J Cell Biol.* 160:699-708.

Ghosh, P., and S. Kornfeld. 2003b. Phosphorylation-induced conformational changes regulate GGAs 1 and 3 function at the trans-Golgi network. *J Biol Chem.* 278:14543-9.

Gillooly, D.J., I.C. Morrow, M. Lindsay, R. Gould, N.J. Bryant, J.M. Gaullier, R.G. Parton, and H. Stenmark. 2000. Localization of phosphatidylinositol 3-phosphate in yeast and mammalian cells. *Embo J.* 19:4577-4588.

Glickman, J.N., E. Conibear, and B.M. Pearse. 1989. Specificity of binding of clathrin adaptors to signals on the mannose-6-phosphate/insulin-like growth factor II receptor. *Embo J.* 8:1041-7.

Glickman, J.N., and S. Kornfeld. 1993. Mannose 6-phosphate-independent targeting of lysosomal enzymes in I- cell disease B lymphoblasts. *J Cell Biol.* 123:99-108.

Glombik, M.M., A. Kromer, T. Salm, W.B. Huttner, and H.H. Gerdes. 1999. The disulfide-bonded loop of chromogranin B mediates membrane binding and directs sorting from the trans-Golgi network to secretory granules. *Embo J.* 18:1059-70.

Godar, S., V. Horejsi, U.H. Weidle, B.R. Binder, C. Hansmann, and H. Stockinger. 1999. M6P/IGFII-receptor complexes urokinase receptor and plasminogen for activation of transforming growth factor-beta1. *Eur J Immunol.* 29:1004-13.

Godi, A., P. Pertile, R. Meyers, P. Marra, G. Di Tullio, C. Iurisci, A. Luini, D. Corda, and M.A. De Matteis. 1999. ARF mediates recruitment of PtdIns-4-OH kinase-beta and stimulates synthesis of PtdIns(4,5)P2 on the Golgi complex. *Nat Cell Biol.* 1:280-7.

Goldberg, J. 1998. Structural basis for activation of ARF GTPase: mechanisms of guanine nucleotide exchange and GTP-myristoyl switching. *Cell.* 95:237-48.

Gonzalez, L., Jr., and R.H. Scheller. 1999. Regulation of membrane trafficking: structural insights from a Rab/effector complex. *Cell.* 96:755-8.

Gorvel, J.P., P. Chavrier, M. Zerial, and J. Gruenberg. 1991. rab5 controls early endosome fusion in vitro. *Cell.* 64:915-25.

Grabs, D., V.I. Slepnev, S.Y. Zhou, C. David, M. Lynch, L.C. Cantley, and P. Decamilli. 1997. The SH3 domain of amphiphysin binds the proline-rich domain of dynamin at a single site that defines a new SH3 binding consensus sequence. *J Biol Chem.* 272:13419-13425.

References

Griffiths, G., S. Pfeiffer, K. Simons, and K. Matlin. 1985. Exit of newly synthesized membrane proteins from the trans cisterna of the Golgi complex to the plasma membrane. *J Cell Biol.* 101:949-64.

Gruenberg, J. 2001. The endocytic pathway: a mosaic of domains. *Nat Rev Mol Cell Biol.* 2:721-30.

Guo, S., L.E. Stolz, S.M. Lemrow, and J.D. York. 1999. SAC1-like domains of yeast SAC1, INP52, and INP53 and of human synaptojanin encode polyphosphoinositide phosphatases. *J Biol Chem.* 274:12990-5.

Gutierrez, L., and A.I. Magee. 1991. Characterization of an acyltransferase acting on p21N-ras protein in a cell-free system. *Biochim Biophys Acta.* 1078:147-54.

Haft, C.R., M. de la Luz Sierra, R. Bafford, M.A. Lesniak, V.A. Barr, and S.I. Taylor. 2000. Human orthologs of yeast vacuolar protein sorting proteins Vps26, 29, and 35: assembly into multimeric complexes. *Mol Biol Cell.* 11:4105-16.

Hancock, J.F., A.I. Magee, J.E. Childs, and C.J. Marshall. 1989. All ras proteins are polyisoprenylated but only some are palmitoylated. *Cell.* 57:1167-77.

Hancock, M.K., D.J. Haskins, G. Sun, and N.M. Dahms. 2002. Identification of residues essential for carbohydrate recognition by the insulin-like growth factor II/mannose 6-phosphate receptor. *J Biol Chem.* 277:11255-64.

Hao, W.H., Z. Luo, L. Zheng, K. Prasad, and E.M. Lafer. 1999. AP180 and AP-2 interact directly in a complex that cooperatively assembles clathrin. *J Biol Chem.* 274:22785-22794.

Haun, R.S., S.C. Tsai, R. Adamik, J. Moss, and M. Vaughan. 1993. Effect of myristoylation on GTP-dependent binding of ADP-ribosylation factor to Golgi. *J Biol Chem.* 268:7064-8.

He, X., W.P. Chang, G. Koelsch, and J. Tang. 2002. Memapsin 2 (beta-secretase) cytosolic domain binds to the VHS domains of GGA1 and GGA2: implications on the endocytosis mechanism of memapsin 2. *FEBS Lett.* 524:183-7.

Hemer, F., C. Körner, and T. Braulke. 1993. Phosphorylation of the human 46-kDa mannose 6-phosphate receptor in the cytoplasmic domain at serine 56. *J Biol Chem.* 268:17108-13.

Hermansky, F., and P. Pudlak. 1959. Albinism associated with hemorrhagic diathesis and unusual pigmented reticular cells in the bone marrow: report of two cases with histochemical studies. *Blood.* 14:162-9.

Hermida-Matsumoto, L., and M.D. Resh. 1999. Human immunodeficiency virus type 1 protease triggers a myristoyl switch that modulates membrane binding of Pr55(gag) and p17MA. *J Virol.* 73:1902-8.

Heuser, J. 1980. Three-dimensional visualization of coated vesicle formation in fibroblasts. *J Cell Biol.* 84:560-83.

Hicke, L. 2001. A New Ticket for Entry into Budding Vesicles-Ubiquitin. *Cell.* 106:527-30.

Hille, A., A. Waheed, and K. von Figura. 1990. Assembly of the ligand-binding conformation of Mr 46,000 mannose 6-phosphate-specific receptor takes place before reaching the Golgi complex. *J Cell Biol.* 110:963-72.

Hille-Rehfeld, A. 1995. Mannose 6-phosphate receptors in sorting and transport of lysosomal enzymes. *Biochim Biophys Acta.* 1241:177-94.

Hinners, I., F. Wendler, H. Fei, L. Thomas, G. Thomas, and S.A. Tooze. 2003. AP-1 recruitment to VAMP4 is modulated by phosphorylation-dependent binding of PACS-1. *EMBO Rep.*

Hinshaw, J.E., and S.L. Schmid. 1995. Dynamin self-assembles into rings suggesting a mechanism for coated vesicle budding. *Nature.* 374:190-2.

Hirst, J., N.A. Bright, B. Rous, and M.S. Robinson. 1999. Characterization of a fourth adaptor-related protein complex. *Mol Biol Cell.* 10:2787-2802.

Hirst, J., W.W.Y. Lui, N.A. Bright, N. Totty, M.N.J. Seaman, and M.S. Robinson. 2000. A family of proteins with gamma-adaptin and VHS domains that facilitate trafficking between the trans-Golgi network and the vacuole/lysosome. *J Cell Biol.* 149:67-79.

Hirst, J., A. Motley, K. Harasaki, S.Y. Peak Chew, and M.S. Robinson. 2003. EpsinR: an ENTH domain-containing protein that interacts with AP-1. *Mol Biol Cell.* 14:625-41.

References

Ho, S.N., H.D. Hunt, R.M. Horton, J.K. Pullen, and L.R. Pease. 1989. Site-directed mutagenesis by overlap extension using the polymerase chain reaction [see comments]. *Gene*. 77:51-9.

Hoflack, B., K. Fujimoto, and S. Kornfeld. 1987. The interaction of phosphorylated oligosaccharides and lysosomal enzymes with bovine liver cation-dependent mannose 6-phosphate receptor. *J Biol Chem*. 262:123-9.

Hoflack, B., and S. Kornfeld. 1985. Purification and characterization of a cation-dependent mannose 6- phosphate receptor from murine P388D1 macrophages and bovine liver. *J Biol Chem*. 260:12008-14.

Höning, S., J. Griffith, H.J. Geuze, and W. Hunziker. 1996. The tyrosine-based lysosomal targeting signal in lamp-1 mediates sorting into Golgi-derived clathrin-coated vesicles. *Embo J*. 15:5230-9.

Höning, S., I.V. Sandoval, and K. Vonfigura. 1998. A di-leucine-based motif in the cytoplasmic tail of LIMP- II and tyrosinase mediates selective binding of AP-3. *Embo J*. 17:1304-1314.

Höning, S., M. Sosa, A. Hille-Rehfeld, and K. von Figura. 1997. The 46-kDa mannose 6-phosphate receptor contains multiple binding sites for clathrin adaptors. *J Biol Chem*. 272:19884-19890.

Horiuchi, H., R. Lippe, H.M. McBride, M. Rubino, P. Woodman, H. Stenmark, V. Rybin, M. Wilm, K. Ashman, M. Mann, and M. Zerial. 1997. A novel Rab5 GDP/GTP exchange factor complexed to Rabaptin- 5 links nucleotide exchange to effector recruitment and function. *Cell*. 90:1149-1159.

Hume, A.N., L.M. Collinson, C.R. Hopkins, M. Strom, D.C. Barral, G. Bossi, G.M. Griffiths, and M.C. Seabra. 2002. The leaden gene product is required with Rab27a to recruit myosin Va to melanosomes in melanocytes. *Traffic*. 3:193-202.

Iiri, T., P.S. Backlund, Jr., T.L. Jones, P.B. Wedegaertner, and H.R. Bourne. 1996. Reciprocal regulation of Gs alpha by palmitate and the beta gamma subunit. *Proc Natl Acad Sci U S A*. 93:14592-7.

Jahn, R. 2000. Sec1/Munc18 proteins: mediators of membrane fusion moving to center stage. *Neuron*. 27:201-4.

Jha, A., N.R. Agostinelli, S.K. Mishra, P.A. Keyel, M.J. Hawryluk, and L.M. Traub. 2003. A novel AP-2 adaptor interaction motif initially identified in the long-splice isoform of synaptojanin 1, SJ170a. *J Biol Chem*.

Johnson, K.F., W. Chan, and S. Kornfeld. 1990. Cation-dependent mannose 6-phosphate receptor contains two internalization signals in its cytoplasmic domain (published erratum appears in Proc Natl Acad Sci U S A 1991 Feb 15; 88(4):1591). *Proc Natl Acad Sci U S A*. 87:10010-4.

Johnson, K.F., and S. Kornfeld. 1992a. The cytoplasmic tail of the mannose 6-phosphate/insulin-like growth factor-II receptor has two signals for lysosomal enzyme sorting in the Golgi. *J Cell Biol*. 119:249-57.

Johnson, K.F., and S. Kornfeld. 1992b. A His-Leu-Leu sequence near the carboxyl terminus of the cytoplasmic domain of the cation-dependent mannose 6-phosphate receptor is necessary for the lysosomal enzyme sorting function. *J Biol Chem*. 267:17110-5.

Jones, S.M., K.E. Howell, J.R. Henley, H. Cao, and M.A. McNiven. 1998. Role of Dynamin in the Formation of Transport Vesicles from the Trans-Golgi Network. *Science*. 279:573-577.

Jones, T.L., M.Y. Degtyarev, and P.S. Backlund, Jr. 1997. The stoichiometry of G alpha(s) palmitoylation in its basal and activated states. *Biochemistry*. 36:7185-91.

Kabouridis, P.S., A.I. Magee, and S.C. Ley. 1997. S-acylation of LCK protein tyrosine kinase is essential for its signalling function in T lymphocytes. *Embo J*. 16:4983-98.

Kalthoff, C., J. Alves, C. Urbanke, R. Knorr, and E.J. Ungewickell. 2002. Unusual structural organization of the endocytic proteins AP180 and epsin 1. *J Biol Chem*. 277:8209-16.

Kanai, F., H. Liu, S.J. Field, H. Akbary, T. Matsuo, G.E. Brown, L.C. Cantley, and M.B. Yaffe. 2001. The PX domains of p47phox and p40phox bind to lipid products of PI(3)K. *Nat Cell Biol*. 3:675-8.

References

Kantheti, P., X.X. Qiao, M.E. Diaz, A.A. Peden, G.E. Meyer, S.L. Carskadon, D. Kapfhamer, D. Sufalko, M.S. Robinson, J.L. Noebels, and M. Burmeister. 1998. Mutation in AP-3 delta in the mocha mouse links endosomal transport to storage deficiency in platelets, melanosomes, and synaptic vesicles. *Neuron.* 21:111-122.

Karnik, S.S., K.D. Ridge, S. Bhattacharya, and H.G. Khorana. 1993. Palmitoylation of bovine opsin and its cysteine mutants in COS cells. *Proc Natl Acad Sci U S A.* 90:40-4.

Kasper, D., F. Dittmer, K. von Figura, and R. Pohlmann. 1996. Neither type of mannose 6-phosphate receptor is sufficient for targeting of lysosomal enzymes along intracellular routes. *J Cell Biol.* 134:615-23.

Kato, Y., S. Misra, R. Puertollano, J.H. Hurley, and J.S. Bonifacino. 2002. Phosphoregulation of sorting signal-VHS domain interactions by a direct electrostatic mechanism. *Nat Struct Biol.* 9:532-6.

Katzmann, D.J., G. Odorizzi, and S.D. Emr. 2002. Receptor downregulation and multivesicular-body sorting. *Nat Rev Mol Cell Biol.* 3:893-905.

Kent, H.M., H.T. McMahon, P.R. Evans, A. Benmerah, and D.J. Owen. 2002. Gamma-adaptin appendage domain: structure and binding site for Eps15 and gamma-synergin. *Structure (Camb).* 10:1139-48.

Kiess, W., L.A. Greenstein, L. Lee, C. Thomas, and S.P. Nissley. 1991. Biosynthesis of the insulin-like growth factor-II (IGF-II)/mannose-6-phosphate receptor in rat C6 glial cells: the role of N-linked glycosylation in binding of IGF-II to the receptor. *Mol Endocrinol.* 5:281-91.

Kim, T., J.H. Tao-Cheng, L.E. Eiden, and Y.P. Loh. 2001. Chromogranin A, an "on/off" switch controlling dense-core secretory granule biogenesis. *Cell.* 106:499-509.

Kirchhausen, T. 2000. Clathrin. *Annu Rev Biochem.* 69:699-727.

Kleuss, C., and E. Krause. 2003. Galpha(s) is palmitoylated at the N-terminal glycine. *Embo J.* 22:826-32.

Klumperman, J., A. Hille, T. Veenendaal, V. Oorschot, W. Stoorvogel, K. von Figura, and H.J. Geuze. 1993. Differences in the endosomal distributions of the two mannose 6- phosphate receptors. *J Cell Biol.* 121:997-1010.

Klumperman, J., R. Kuliawat, J.M. Griffith, H.J. Geuze, and P. Arvan. 1998. Mannose 6-phosphate receptors are sorted from immature secretory granules via adaptor protein AP-1, clathrin, and syntaxin 6-positive vesicles. *J Cell Biol.* 141:359-71.

Kobayashi, T., E. Stang, K.S. Fang, P. Demoerloose, R.G. Parton, and J. Gruenberg. 1998. A lipid associated with the antiphospholipid syndrome regulates endosome structure and function. *Nature.* 392:193-197.

Koegl, M., P. Zlatkine, S.C. Ley, S.A. Courtneidge, and A.I. Magee. 1994. Palmitoylation of multiple Src-family kinases at a homologous N-terminal motif. *Biochem J.* 303:749-53.

Komada, M., K. Hatsuzawa, S. Shibamoto, F. Ito, K. Nakayama, and N. Kitamura. 1993. Proteolytic processing of the hepatocyte growth factor/scatter factor receptor by furin. *FEBS Lett.* 328:25-9.

Komada, M., and N. Kitamura. 1995. Growth factor-induced tyrosine phosphorylation of Hrs, a novel 115-kilodalton protein with a structurally conserved putative zinc finger domain. *Mol Cell Biol.* 15:6213-21.

Körner, C., A. Herzog, B. Weber, O. Rosorius, F. Hemer, B. Schmidt, and T. Braulke. 1994. In vitro phosphorylation of the 46-kDa mannose 6-phosphate receptor by casein kinase II. Structural requirements for efficient phosphorylation. *J Biol Chem.* 269:16529-32.

Kornfeld, R., M. Bao, K. Brewer, C. Noll, and W.M. Canfield. 1998. Purification and multimeric structure of bovine N- acetylglucosamine-1-phosphodiester alpha-N- acetylglucosaminidase. *J Biol Chem.* 273:23203-23210.

Kornfeld, S. 1992. Structure and function of the mannose 6-phosphate/insulinlike growth factor II receptors. *Annu Rev Biochem.* 61:307-30.

Kornfeld, S., and I. Mellman. 1989. The biogenesis of lysosomes. *Annu Rev Cell Biol.* 5:483-525.

Kozlov, M.M. 2001. Fission of biological membranes: Interplay between dynamin and lipids. *Traffic*. 2:51-65.

Kreitzer, G., A. Marmorstein, P. Okamoto, R. Vallee, and E. Rodriguez-Boulan. 2000. Kinesin and dynamin are required for post-Golgi transport of a plasma-membrane protein. *Nat Cell Biol*. 2:125-7.

Kundra, A., and S. Kornfeld. 1998. Wortmannin retards the movement of the mannose 6-phosphate/insulin-like growth factor II receptor and its ligand out of endosomes. *J Biol Chem*. 273:3848-3853.

Kundra, R., and S. Kornfeld. 1999. Asparagine-linked oligosaccharides protect lamp-1 and lamp-2 from intracellular proteolysis [In Process Citation]. *J Biol Chem*. 274:31039-46.

Kurten, R.C., D.L. Cadena, and G.N. Gill. 1996. Enhanced degradation of EGF receptors by a sorting nexin, SNX1. *Science*. 272:1008-10.

Laemmli, U.K. 1970. Cleavage of structural proteins during the assembly of the head of bacteriophage T4. *Nature*. 227:680-5.

Laporte, S.A., W.E. Miller, K.M. Kim, and M.G. Caron. 2002. beta-Arrestin/AP-2 interaction in G protein-coupled receptor internalization: identification of a beta-arrestin binging site in beta 2-adaptin. *J Biol Chem*. 277:9247-54.

Le Borgne, R., A. Alconada, U. Bauer, and B. Hoflack. 1998. The mammalian AP-3 adaptor-like complex mediates the intracellular transport of lysosomal membrane glycoproteins. *J Biol Chem*. 273:29451-61.

Le Borgne, R., and B. Hoflack. 1997. Mannose 6-phosphate receptors regulate the formation of clathrin-coated vesicles in the TGN. *J Cell Biol*. 137:335-45.

Le Borgne, R., A. Schmidt, F. Mauxion, G. Griffiths, and B. Hoflack. 1993. Binding of AP-1 Golgi adaptors to membranes requires phosphorylated cytoplasmic domains of the mannose 6-phosphate/insulin-like growth factor II receptor. *J Biol Chem*. 268:22552-6.

Legendre-Guillemin, V., M. Metzler, M. Charbonneau, L. Gan, V. Chopra, J. Philie, M.R. Hayden, and P.S. McPherson. 2002. HIP1 and HIP12 display differential binding to F-actin, AP2, and clathrin. Identification of a novel interaction with clathrin light chain. *J Biol Chem*. 277:19897-904.

Leventis, R., G. Juel, J.K. Knudsen, and J.R. Silvius. 1997. Acyl-CoA binding proteins inhibit the nonenzymic S-acylation of cysteinyl-containing peptide sequences by long-chain acyl-CoAs. *Biochemistry*. 36:5546-53.

Li, M., J.J. Distler, and G.W. Jourdian. 1990. The aggregation and dissociation properties of a low molecular weight mannose 6-phosphate receptor from bovine testis. *Arch Biochem Biophys*. 283:150-7.

Lippincott-Schwartz, J., and D.M. Fambrough. 1987. Cycling of the integral membrane glycoprotein, LEP100, between plasma membrane and lysosomes: kinetic and morphological analysis. *Cell*. 49:669-77.

Liu, L., T. Dudler, and M.H. Gelb. 1996. Purification of a protein palmitoyltransferase that acts on H-Ras protein and on a C-terminal N-Ras peptide [published erratum appears in J Biol Chem 1999 Jan 29;274(5):3252]. *J Biol Chem*. 271:23269-76.

Liu, L., T. Dudler, and M.H. Gelb. 1999. Purification of a protein palmitoyltransferase that acts on H-Ras protein and on a C-terminal N-Ras peptide. (Vol 271, pg 23269, 1996). *J Biol Chem*. 274:3252.

Lobel, P., N.M. Dahms, and S. Kornfeld. 1988. Cloning and sequence analysis of the cation-independent mannose 6- phosphate receptor. *J Biol Chem*. 263:2563-70.

Lobo, S., W.K. Greentree, M.E. Linder, and R.J. Deschenes. 2002. Identification of a Ras palmitoyltransferase in Saccharomyces cerevisiae. *J Biol Chem*. 277:41268-73.

Loisel, T.P., L. Adam, T.E. Hebert, and M. Bouvier. 1996. Agonist stimulation increases the turnover rate of beta 2AR-bound palmitate and promotes receptor depalmitoylation. *Biochemistry*. 35:15923-32.

References

Lu, J.Y., L.A. Verkruyse, and S.L. Hofmann. 1996. Lipid thioesters derived from acylated proteins accumulate in infantile neuronal ceroid lipofuscinosis: correction of the defect in lymphoblasts by recombinant palmitoyl-protein thioesterase. *Proc Natl Acad Sci U S A*. 93:10046-50.

Ludwig, T., H. Munier-Lehmann, U. Bauer, M. Hollinshead, C. Ovitt, P. Lobel, and B. Hoflack. 1994. Differential sorting of lysosomal enzymes in mannose 6-phosphate receptor-deficient fibroblasts. *Embo J*. 13:3430-7.

Ludwig, T., C.E. Ovitt, U. Bauer, M. Hollinshead, J. Remmler, P. Lobel, U. Ruther, and B. Hoflack. 1993. Targeted disruption of the mouse cation-dependent mannose 6-phosphate receptor results in partial missorting of multiple lysosomal enzymes. *Embo J*. 12:5225-35.

Lui, W.W., B.M. Collins, J. Hirst, A. Motley, C. Millar, P. Schu, D.J. Owen, and M.S. Robinson. 2003. Binding partners for the COOH-terminal appendage domains of the GGAs and gamma-adaptin. *Mol Biol Cell*. 14:2385-98.

Luzio, J.P., B.A. Rous, N.A. Bright, P.R. Pryor, B.M. Mullock, and R.C. Piper. 2000. Lysosome-endosome fusion and lysosome biogenesis. *J Cell Sci*. 113:1515-1524.

Machen, T.E., M.J. Leigh, C. Taylor, T. Kimura, S. Asano, and H.P. Moore. 2003. pH of TGN and recycling endosomes of H+/K+-ATPase-transfected HEK-293 cells: implications for pH regulation in the secretory pathway. *Am J Physiol Cell Physiol*. 285:C205-14.

Mallet, W.G., and F.R. Maxfield. 1999. Chimeric forms of furin and TGN38 are transported from the plasma membrane to the trans-Golgi network via distinct endosomal pathways. *J Cell Biol*. 146:345-359.

Mansour, S.J., J. Skaug, X.H. Zhao, J. Giordano, S.W. Scherer, and P. Melancon. 1999. p200 ARF-GEP1: a Golgi-localized guanine nucleotide exchange protein whose Sec7 domain is targeted by the drug brefeldin A. *Proc Natl Acad Sci U S A*. 96:7968-73.

Marks, M.S., L. Woodruff, H. Ohno, and J.S. Bonifacino. 1996. Protein targeting by tyrosine- and di-leucine-based signals: evidence for distinct saturable components. *J Cell Biol*. 135:341-54.

Marron-Terada, P.G., K.E. Bollinger, and N.M. Dahms. 1998a. Characterization of truncated and glycosylation-deficient forms of the cation-dependent mannose 6-phosphate receptor expressed in baculovirus-infected insect cells. *Biochemistry-Usa*. 37:17223-17229.

Marron-Terada, P.G., M.A. Brzyckiwessell, and N.M. Dahms. 1998b. The two mannose 6-phosphate binding sites of the insulin- like growth factor-II/mannose 6-phosphate receptor display different ligand binding properties. *J Biol Chem*. 273:22358-22366.

Marron-Terada, P.G., M.K. Hancock, D.J. Haskins, and N.M. Dahms. 2000. Recognition of Dictyostelium discoideum lysosomal enzymes is conferred by the amino-terminal carbohydrate binding site of the insulin-like growth factor II/mannose 6- phosphate receptor. *Biochemistry Usa*. 39:2243-2253.

Marsh, M., G. Griffiths, G.E. Dean, I. Mellman, and A. Helenius. 1986. Three-dimensional structure of endosomes in BHK-21 cells. *Proc Natl Acad Sci U S A*. 83:2899-903.

Marsh, M., and H.T. McMahon. 1999. Cell biology - The structural era of endocytosis. *Science*. 285:215-220.

Marxer, A., B. Stieger, A. Quaroni, M. Kashgarian, and H.P. Hauri. 1989. (Na+ + K+)-ATPase and plasma membrane polarity of intestinal epithelial cells: presence of a brush border antigen in the distal large intestine that is immunologically related to beta subunit. *J Cell Biol*. 109:1057-69.

Matlin, K.S., and K. Simons. 1983. Reduced temperature prevents transfer of a membrane glycoprotein to the cell surface but does not prevent terminal glycosylation. *Cell*. 34:233-43.

Matzner, U., K. von Figura, and R. Pohlmann. 1992. Expression of the two mannose 6-phosphate receptors is spatially and temporally different during mouse embryogenesis. *Development*. 114:965-72.

Maurer-Stroh, S., S. Washietl, and F. Eisenhaber. 2003. Protein prenyltransferases. *Genome Biol*. 4:212.

Mauxion, F., R. Le Borgne, H. Munier-Lehmann, and B. Hoflack. 1996. A casein kinase II phosphorylation site in the cytoplasmic domain of the cation-dependent mannose 6-phosphate receptor determines the high affinity interaction of the AP-1 Golgi assembly proteins with membranes. *J Biol Chem*. 271:2171-8.

McCabe, J.B., and L.G. Berthiaume. 1999. Functional roles for fatty acylated amino-terminal domains in subcellular localization. *Mol Biol Cell*. 10:3771-86.

McCabe, J.B., and L.G. Berthiaume. 2001. N-terminal protein acylation confers localization to cholesterol, sphingolipid-enriched membranes but not to lipid rafts/caveolae. *Mol Biol Cell*. 12:3601-17.

McGill, M.A., and C.J. McGlade. 2003. Mammalian numb proteins promote Notch1 receptor ubiquitination and degradation of the Notch1 intracellular domain. *J Biol Chem*. 278:23196-203.

McLauchlan, H., J. Newell, N. Morrice, A. Osborne, M. West, and E. Smythe. 1998. A novel role for Rab5-GDI in ligand sequestration into calthrin-coated pits. *Curr Biol*. 8:34-45.

McLaughlin, S., and A. Aderem. 1995. The myristoyl-electrostatic switch: a modulator of reversible protein-membrane interactions. *Trends Biochem Sci*. 20:272-6.

McMahon, H.T., P. Wigge, and C. Smith. 1997. Clathrin interacts specifically with amphiphysin and is displaced by dynamin. *FEBS Lett*. 413:319-322.

McPherson, P.S., E.P. Garcia, V.I. Slepnev, C. David, X. Zhang, D. Grabs, W.S. Sossin, R. Bauerfeind, Y. Nemoto, and P. De Camilli. 1996. A presynaptic inositol-5-phosphatase. *Nature*. 379:353-7.

Medigeshi, G.R., and P. Schu. 2003. Characterization of the in Vitro Retrograde Transport of MPR46. *Traffic*. 4:802-811.

Meggio, F., and L.A. Pinna. 2003. One-thousand-and-one substrates of protein kinase CK2? *Faseb J*. 17:349-68.

Menasche, G., E. Pastural, J. Feldmann, S. Certain, F. Ersoy, S. Dupuis, N. Wulffraat, D. Bianchi, A. Fischer, F. Le Deist, and G. de Saint Basile. 2000. Mutations in RAB27A cause Griscelli syndrome associated with haemophagocytic syndrome. *Nat Genet*. 25:173-6.

Meresse, S., and B. Hoflack. 1993. Phosphorylation of the cation-independent mannose 6-phosphate receptor is closely associated with its exit from the trans-Golgi network. *J Cell Biol*. 120:67-75.

Meresse, S., T. Ludwig, R. Frank, and B. Hoflack. 1990. Phosphorylation of the cytoplasmic domain of the bovine cation-independent mannose 6-phosphate receptor. Serines 2421 and 2492 are the targets of a casein kinase II associated to the Golgi-derived HAI adaptor complex. *J Biol Chem*. 265:18833-42.

Messner, D.J. 1993. The mannose receptor and the cation-dependent form of mannose 6- phosphate receptor have overlapping cellular and subcellular distributions in liver. *Arch Biochem Biophys*. 306:391-401.

Metzler, M., V. Legendre-Guillemin, L. Gan, V. Chopra, A. Kwok, P.S. McPherson, and M.R. Hayden. 2001. HIP1 functions in clathrin-mediated endocytosis through binding to clathrin and adaptor protein 2. *J Biol Chem*. 276:39271-6.

Meyer, C., D. Zizioli, S. Lausmann, E.L. Eskelinen, J. Hamann, P. Saftig, K. von Figura, and P. Schu. 2000. mu 1A-adaptin-deficient mice: lethality, loss of AP-1 binding and rerouting of mannose 6-phosphate receptors. *Embo J*. 19:2193-2203.

Miller, W.E., and R.J. Lefkowitz. 2001. Expanding roles for beta-arrestins as scaffolds and adapters in GPCR signaling and trafficking. *Curr Opin Cell Biol*. 13:139-45.

Milligan, G., M. Parenti, and A.I. Magee. 1995. The dynamic role of palmitoylation in signal transduction. *Trends Biochem Sci*. 20:181-7.

Mills, I.G., G.J. Praefcke, Y. Vallis, B.J. Peter, L.E. Olesen, J.L. Gallop, P.J. Butler, P.R. Evans, and H.T. McMahon. 2003. EpsinR: an AP1/clathrin interacting protein involved in vesicle trafficking. *J Cell Biol*. 160:213-22.

Mishra, S.K., P.A. Keyel, M.J. Hawryluk, N.R. Agostinelli, S.C. Watkins, and L.M. Traub. 2002. Disabled-2 exhibits the properties of a cargo-selective endocytic clathrin adaptor. *Embo J.* 21:4915-26.

Misra, S., R. Puertollano, Y. Kato, J.S. Bonifacino, and J.H. Hurley. 2002. Structural basis for acidic-cluster-dileucine sorting-signal recognition by VHS domains. *Nature.* 415:933-7.

Mizuno, E., K. Kawahata, M. Kato, N. Kitamura, and M. Komada. 2003. STAM proteins bind ubiquitinated proteins on the early endosome via the VHS domain and ubiquitin-interacting motif. *Mol Biol Cell.* 14:3675-89.

Moffett, S., L. Adam, H. Bonin, T.P. Loisel, M. Bouvier, and B. Mouillac. 1996. Palmitoylated cysteine 341 modulates phosphorylation of the beta2-adrenergic receptor by the cAMP-dependent protein kinase. *J Biol Chem.* 271:21490-7.

Moffett, S., G. Rousseau, M. Lagace, and M. Bouvier. 2001. The palmitoylation state of the beta(2)-adrenergic receptor regulates the synergistic action of cyclic AMP-dependent protein kinase and beta-adrenergic receptor kinase involved in its phosphorylation and desensitization. *J Neurochem.* 76:269-79.

Morgan, A., R. Dimaline, and R.D. Burgoyne. 1994. The ATPase activity of N-ethylmaleimide-sensitive fusion protein (NSF) is regulated by soluble NSF attachment proteins. *J Biol Chem.* 269:29347-50.

Morgan, J.R., K. Prasad, W.H. Hao, G.J. Augustine, and E.M. Lafer. 2000. A conserved clathrin assembly motif essential for synaptic vesicle endocytosis. *J Neurosci.* 20:8667-8676.

Morgan, J.R., K. Prasad, S. Jin, G.J. Augustine, and E.M. Lafer. 2001. Uncoating of clathrin-coated vesicles in presynaptic terminals: roles for Hsc70 and auxilin. *Neuron.* 32:289-300.

Morris, S.M., and J.A. Cooper. 2001. Disabled-2 colocalizes with the LDLR in clathrin-coated pits and interacts with AP-2. *Traffic.* 2:111-23.

Muhlberg, A.B., D.E. Warnock, and S.L. Schmid. 1997. Domain structure and intramolecular regulation of dynamin GTPase. *Embo J.* 16:6676-6683.

Mullins, C., and J.S. Bonifacino. 2001. The molecular machinery for lysosome biogenesis. *Bioessays.* 23:333-43.

Mundy, D.I., and G. Warren. 1992. Mitosis and inhibition of intracellular transport stimulate palmitoylation of a 62-kD protein. *J Cell Biol.* 116:135-46.

Nair, P., B.E. Schaub, and J. Rohrer. 2003. Characterization of the endosomal sorting signal of the cation-dependent mannose 6-phosphate receptor. *J Biol Chem.* 278:24753-8.

Nguyen, D.H., and J.E. Hildreth. 2000. Evidence for budding of human immunodeficiency virus type 1 selectively from glycolipid-enriched membrane lipid rafts. *J Virol.* 74:3264-72.

Nielsen, E., S. Christoforidis, S. UttenweilerJoseph, M. Miaczynska, F. Dewitte, M. Wilm, B. Hoflack, and M. Zerial. 2000. Rabenosyn-5, a novel Rab5 effector, is complexed with hVPS45 and recruited to endosomes through a FYVE finger domain. *J Cell Biol.* 151:601-612.

Nielsen, E., F. Severin, J.M. Backer, A.A. Hyman, and M. Zerial. 1999. Rab5 regulates motility of early endosomes on microtubules. *Nat Cell Biol.* 1:376-382.

O'Brien, P.J., R.S. St Jules, T.S. Reddy, N.G. Bazan, and M. Zatz. 1987. Acylation of disc membrane rhodopsin may be nonenzymatic. *J Biol Chem.* 262:5210-5.

O'Dell, S.D., and I.N. Day. 1998. Insulin-like growth factor II (IGF-II). *Int J Biochem Cell Biol.* 30:767-71.

Ohno, H., R.C. Aguilar, D. Yeh, D. Taura, T. Saito, and J.S. Bonifacino. 1998. The medium subunits of adaptor complexes recognize distinct but overlapping sets of tyrosine-based sorting signals. *J Biol Chem.* 273:25915-25921.

Ohno, H., M.C. Fournier, G. Poy, and J.S. Bonifacino. 1996. Structural determinants of interaction of tyrosine-based sorting signals with the adaptor medium chains. *J Biol Chem.* 271:29009-15.

Ohno, H., J. Stewart, M.C. Fournier, H. Bosshart, I. Rhee, S. Miyatake, T. Saito, A. Gallusser, T. Kirchhausen, and J.S. Bonifacino. 1995. Interaction of tyrosine-based sorting signals with clathrin-associated proteins. *Science.* 269:1872-5.

Ohno, H., T. Tomemori, F. Nakatsu, Y. Okazaki, R.C. Aguilar, H. Foelsch, I. Mellman, T. Saito, T. Shirasawa, and J.S. Bonifacino. 1999. Mu1B, a novel adaptor medium chain expressed in polarized epithelial cells. *FEBS Lett.* 449:215-20.

Ohshima, T., G.J. Murray, W.D. Swaim, G. Longenecker, J.M. Quirk, C.O. Cardarelli, Y. Sugimoto, I. Pastan, M.M. Gottesman, R.O. Brady, and A.B. Kulkarni. 1997. alpha-Galactosidase A deficient mice: a model of Fabry disease. *Proc Natl Acad Sci U S A.* 94:2540-4.

Olkkonen, V.M., and E. Ikonen. 2000. Genetic defects of intracellular-membrane transport. *N Engl J Med.* 343:1095-104.

Olson, L.J., J. Zhang, N.M. Dahms, and J.J. Kim. 2002. Twists and turns of the cation-dependent mannose 6-phosphate receptor. Ligand-bound versus ligand-free receptor. *J Biol Chem.* 277:10156-61.

Olson, L.J., J. Zhang, Y.C. Lee, N.M. Dahms, and J.J.P. Kim. 1999. Structural basis for recognition of phosphorylated high mannose oligosaccharides by the cation-dependent mannose 6-phosphate receptor. *J Biol Chem.* 274:29889-29896.

Ooi, C.E., E.C. Dellangelica, and J.S. Bonifacino. 1998a. ADP-ribosylation factor 1 (ARF1) regulates recruitment of the AP-3 adaptor complex to membranes. *J Cell Biol.* 142:391-402.

Ooi, C.E., E.C. Dell'Angelica, and J.S. Bonifacino. 1998b. ADP-Ribosylation factor 1 (ARF1) regulates recruitment of the AP-3 adaptor complex to membranes. *J Cell Biol.* 142:391-402.

Orsel, J.G., P.M. Sincock, J.P. Krise, and S.R. Pfeffer. 2000. Recognition of the 300-kDa mannose 6-phosphate receptor cytoplasmic domain by 47-kDa tail-interacting protein. *Proc Nat Acad Sci Usa.* 97:9047-9051.

Ostermann, J., L. Orci, K. Tani, M. Amherdt, M. Ravazzola, Z. Elazar, and J.E. Rothman. 1993. Stepwise assembly of functionally active transport vesicles. *Cell.* 75:1015-25.

Owen, D.J., and P.R. Evans. 1998. A structural explanation for the recognition of tyrosine-based endocytotic signals. *Science.* 282:1327-32.

Owen, D.J., Y. Vallis, M.E.M. Noble, J.B. Hunter, T.R. Dafforn, P.R. Evans, and H.T. McMahon. 1999. A structural explanation for the binding of multiple ligands by the alpha-adaptin appendage domain. *Cell.* 97:805-815.

Owen, D.J., Y. Vallis, B.M.F. Pearse, H.T. McMahon, and P.R. Evans. 2000. The structure and function of the beta 2-adaptin appendage domain. *Embo J.* 19:4216-4227.

Page, L.J., P.J. Sowerby, W.W.Y. Lui, and M.S. Robinson. 1999. gamma-Synergin: An EH domain-containing protein that interacts with gamma-adaptin. *J Cell Biol.* 146:993-1004.

Pearse, B.M. 1976. Clathrin: a unique protein associated with intracellular transfer of membrane by coated vesicles. *Proc Natl Acad Sci U S A.* 73:1255-9.

Peden, A.A., R.E. Rudge, W.W. Lui, and M.S. Robinson. 2002. Assembly and function of AP-3 complexes in cells expressing mutant subunits. *J Cell Biol.* 156:327-36.

Pepinsky, R.B., C. Zeng, D. Wen, P. Rayhorn, D.P. Baker, K.P. Williams, S.A. Bixler, C.M. Ambrose, E.A. Garber, K. Miatkowski, F.R. Taylor, E.A. Wang, and A. Galdes. 1998. Identification of a palmitic acid-modified form of human Sonic hedgehog. *J Biol Chem.* 273:14037-45.

Phillips, S.A., V.A. Barr, D.H. Haft, S.I. Taylor, and C.R. Haft. 2001. Identification and characterization of SNX15, a novel sorting nexin involved in protein trafficking. *J Biol Chem.* 276:5074-84.

Piper, R.C., and J.P. Luzio. 2001. Late endosomes: sorting and partitioning in multivesicular bodies. *Traffic.* 2:612-21.

Pippig, S., S. Andexinger, and M.J. Lohse. 1995. Sequestration and recycling of beta 2-adrenergic receptors permit receptor resensitization. *Mol Pharmacol.* 47:666-76.

References

Pohlmann, R., M.W. Boeker, and K. von Figura. 1995. The two mannose 6-phosphate receptors transport distinct complements of lysosomal proteins. *J Biol Chem*. 270:27311-8.

Poirier, M.A., W. Xiao, J.C. Macosko, C. Chan, Y.K. Shin, and M.K. Bennett. 1998. The synaptic SNARE complex is a parallel four-stranded helical bundle. *Nat Struct Biol*. 5:765-9.

Polo, S., S. Sigismund, M. Faretta, M. Guidi, M.R. Capua, G. Bossi, H. Chen, P. De Camilli, and P.P. Di Fiore. 2002. A single motif responsible for ubiquitin recognition and monoubiquitination in endocytic proteins. *Nature*. 416:451-5.

Ponimaskin, E., and M.F. Schmidt. 1998. Domain-structure of cytoplasmic border region is main determinant for palmitoylation of influenza virus hemagglutinin (H7). *Virology*. 249:325-35.

Ponting, C.P. 1996. Novel domains in NADPH oxidase subunits, sorting nexins, and PtdIns 3-kinases: binding partners of SH3 domains? *Protein Sci*. 5:2353-7.

Porter, J.A., K.E. Young, and P.A. Beachy. 1996. Cholesterol modification of hedgehog signaling proteins in animal development. *Science*. 274:255-9.

Prekeris, R., J. Klumperman, Y.A. Chen, and R.H. Scheller. 1998. Syntaxin 13 mediates cycling of plasma membrane proteins via tubulovesicular recycling endosomes. *J Cell Biol*. 143:957-971.

Puertollano, R., R.C. Aguilar, I. Gorshkova, R.J. Crouch, and J.S. Bonifacino. 2001a. Sorting of mannose 6-phosphate receptors mediated by the GGAs. *Science*. 292:1712-6.

Puertollano, R., P.A. Randazzo, J.F. Presley, L.M. Hartnell, and J.S. Bonifacino. 2001b. The GGAs promote ARF-dependent recruitment of clathrin to the TGN. *Cell*. 105:93-102.

Puertollano, R., N.N. van der Wel, L.E. Greene, E. Eisenberg, P.J. Peters, and J.S. Bonifacino. 2003. Morphology and dynamics of clathrin/GGA1-coated carriers budding from the trans-Golgi network. *Mol Biol Cell*. 14:1545-57.

Punnonen, E.L., T. Wilke, K. von Figura, and A. Hille-Rehfeld. 1996. The oligomeric state of 46-kDa mannose 6-phosphate receptor does not change upon intracellular recycling and binding of ligands. *Eur J Biochem*. 237:809-18.

Qanbar, R., and M. Bouvier. 2003. Role of palmitoylation/depalmitoylation reactions in G-protein-coupled receptor function. *Pharmacol Ther*. 97:1-33.

Raiborg, C., K.G. Bache, D.J. Gillooly, I.H. Madshus, E. Stang, and H. Stenmark. 2002. Hrs sorts ubiquitinated proteins into clathrin-coated microdomains of early endosomes. *Nat Cell Biol*. 4:394-8.

Raiborg, C., K.G. Bache, A. Mehlum, E. Stang, and H. Stenmark. 2001a. Hrs recruits clathrin to early endosomes. *Embo J*. 20:5008-21.

Raiborg, C., K.G. Bache, A. Mehlum, and H. Stenmark. 2001b. Function of Hrs in endocytic trafficking and signalling. *Biochem Soc Trans*. 29:472-5.

Raiborg, C., T.E. Rusten, and H. Stenmark. 2003. Protein sorting into multivesicular endosomes. *Curr Opin Cell Biol*. 15:446-55.

Ramjaun, A.R., and P.S. McPherson. 1998. Multiple amphiphysin II splice variants display differential clathrin binding: Identification of two distinct clathrin-binding sites. *J Neurochem*. 70:2369-2376.

Ramjaun, A.R., J. Philie, E. de Heuvel, and P.S. McPherson. 1999. The N terminus of amphiphysin II mediates dimerization and plasma membrane targeting. *J Biol Chem*. 274:19785-91.

Rapoport, I., Y.C. Chen, P. Cupers, S.E. Shoelson, and T. Kirchhausen. 1998. Dileucine-based sorting signals bind to the beta chain of AP-1 at a site distinct and regulated differently from the tyrosine-based motif-binding site. *Embo J*. 17:2148-2155.

Resh, M.D. 1999. Fatty acylation of proteins: new insights into membrane targeting of myristoylated and palmitoylated proteins. *Biochim Biophys Acta*. 1451:1-16.

Ricotta, D., S.D. Conner, S.L. Schmid, K. von Figura, and S. Honing. 2002. Phosphorylation of the AP2 mu subunit by AAK1 mediates high affinity binding to membrane protein sorting signals. *J Cell Biol*. 156:791-5.

Ringstad, N., Y. Nemoto, and P. De Camilli. 1997. The SH3p4/Sh3p8/SH3p13 protein family: binding partners for synaptojanin and dynamin via a Grb2-like Src homology 3 domain. *Proc Natl Acad Sci U S A*. 94:8569-74.

Roberts, D.L., D.J. Weix, N.M. Dahms, and J.J.P. Kim. 1998. Molecular basis of lysosomal enzyme recognition: Three- dimensional structure of the cation-dependent mannose 6- phosphate receptor. *Cell*. 93:639-648.

Robinson, M.S., and J.S. Bonifacino. 2001. Adaptor-related proteins. *Curr Opin Cell Biol*. 13:444-53.

Rodionov, D.G., and O. Bakke. 1998. Medium chains of adaptor complexes AP-1 and AP-2 recognize leucine-based sorting signals from the invariant chain. *J Biol Chem*. 273:6005-6008.

Rohrer, J., and R. Kornfeld. 2001. Lysosomal hydrolase mannose 6-phosphate uncovering enzyme resides in the trans-Golgi network. *Mol Biol Cell*. 12:1623-31.

Rohrer, J., A. Schweizer, K.F. Johnson, and S. Kornfeld. 1995. A determinant in the cytoplasmic tail of the cation-dependent mannose 6- phosphate receptor prevents trafficking to lysosomes. *J Cell Biol*. 130:1297-306.

Rohrer, J., A. Schweizer, D. Russell, and S. Kornfeld. 1996. The targeting of Lamp1 to lysosomes is dependent on the spacing of its cytoplasmic tail tyrosine sorting motif relative to the membrane. *J Cell Biol*. 132:565-76.

Rosenthal, J.A., H. Chen, V.I. Slepnev, L. Pellegrini, A.E. Salcini, P.P. Difiore, and P. Decamilli. 1999. The epsins define a family of proteins that interact with components of the clathrin coat and contain a new protein module. *J Biol Chem*. 274:33959-33965.

Roskoski, R., Jr. 2003. Protein prenylation: a pivotal posttranslational process. *Biochem Biophys Res Commun*. 303:1-7.

Roth, A.F., Y. Feng, L. Chen, and N.G. Davis. 2002. The yeast DHHC cysteine-rich domain protein Akr1p is a palmitoyl transferase. *J Cell Biol*. 159:23-8.

Rous, B.A., B.J. Reaves, G. Ihrke, J.A. Briggs, S.R. Gray, D.J. Stephens, G. Banting, and J.P. Luzio. 2002. Role of adaptor complex AP-3 in targeting wild-type and mutated CD63 to lysosomes. *Mol Biol Cell*. 13:1071-82.

Rybin, V., O. Ullrich, M. Rubino, K. Alexandrov, I. Simon, M.C. Seabra, R. Goody, and M. Zerial. 1996. GTPase activity of Rab5 acts as a timer for endocytic membrane fusion [see comments]. *Nature*. 383:266-9.

Sachse, M., S. Urbe, V. Oorschot, G.J. Strous, and J. Klumperman. 2002. Bilayered clathrin coats on endosomal vacuoles are involved in protein sorting toward lysosomes. *Mol Biol Cell*. 13:1313-28.

Sahagian, G.G., and E.F. Neufeld. 1983. Biosynthesis and turnover of the mannose 6-phosphate receptor in cultured Chinese hamster ovary cells. *J Biol Chem*. 258:7121-8.

Salcini, A.E., S. Confalonieri, M. Doria, E. Santolini, E. Tassi, O. Minenkova, G. Cesareni, P.G. Pelicci, and P.P. Di Fiore. 1997. Binding specificity and in vivo targets of the EH domain, a novel protein-protein interaction module. *Genes Dev*. 11:2239-49.

Salim, K., M.J. Bottomley, E. Querfurth, M.J. Zvelebil, I. Gout, R. Scaife, R.L. Margolis, R. Gigg, C.I. Smith, P.C. Driscoll, M.D. Waterfield, and G. Panayotou. 1996. Distinct specificity in the recognition of phosphoinositides by the pleckstrin homology domains of dynamin and Bruton's tyrosine kinase. *Embo J*. 15:6241-50.

Sambrook, J., E.F. Fritsch, and T. Maniatis. 1998. Molecular cloning: A laboratory manual. Cold Spring Harbor Laboratory Press, Cold Spring Harbor, NY.

Santolini, E., C. Puri, A.E. Salcini, M.C. Gagliani, P.G. Pelicci, C. Tacchetti, and P.P. Di Fiore. 2000. Numb is an endocytic protein. *J Cell Biol*. 151:1345-52.

Santolini, E., A.E. Salcini, B.K. Kay, M. Yamabhai, and P.P. Di Fiore. 1999. The EH network. *Exp Cell Res*. 253:186-209.

Schekman, R., and L. Orci. 1996. Coat proteins and vesicle budding. *Science*. 271:1526-33.

Schmidt, A., M. Wolde, C. Thiele, W. Fest, H. Kratzin, A.V. Podtelejnikov, W. Witke, W.B. Huttner, and H.D. Soling. 1999. Endophilin I mediates synaptic vesicle formation by transfer of arachidonate to lysophosphatidic acid. *Nature*. 401:133-141.

Schweizer, A., M. Ericsson, T. Bachi, G. Griffiths, and H.P. Hauri. 1993a. Characterization of a novel 63 kDa membrane protein. Implications for the organization of the ER-to-Golgi pathway. *J Cell Sci*. 104:671-83.

Schweizer, A., S. Kornfeld, and J. Rohrer. 1996. Cysteine34 of the cytoplasmic tail of the cation-dependent mannose 6-phosphate receptor is reversibly palmitoylated and required for normal trafficking and lysosomal enzyme sorting. *J Cell Biol*. 132:577-84.

Schweizer, A., S. Kornfeld, and J. Rohrer. 1997. Proper sorting of the cation-dependent mannose 6-phosphate receptor in endosomes depends on a pair of aromatic amino acids in its cytoplasmic tail. *Proc Natl Acad Sci USA*. 94:14471-14476.

Schweizer, A., J. Rohrer, P. Jeno, A. DeMaio, T.G. Buchman, and H.P. Hauri. 1993b. A reversibly palmitoylated resident protein (p63) of an ER-Golgi intermediate compartment is related to a circulatory shock resuscitation protein. *J Cell Sci*. 104:685-94.

Schweizer, A., J. Rohrer, and S. Kornfeld. 1995. Determination of the structural requirements for palmitoylation of p63. *J Biol Chem*. 270:9638-44.

Seaman, M.N.J., J.M. McCaffery, and S.D. Emr. 1998. A membrane coat complex essential for endosome-to-Golgi retrograde transport in yeast. *J Cell Biol*. 142:665-681.

Sengar, A.S., W. Wang, J. Bishay, S. Cohen, and S.E. Egan. 1999. The EH and SH3 domain Ese proteins regulate endocytosis by linking to dynamin and Eps15. *Embo J*. 18:1159-1171.

Sever, S., A.B. Muhlberg, and S.L. Schmid. 1999. Impairment of dynamin's GAP domain stimulates receptor- mediated endocytosis. *Nature*. 398:481-486.

Shenoy, S.K., P.H. McDonald, T.A. Kohout, and R.J. Lefkowitz. 2001. Regulation of Receptor Fate by Ubiquitination of Activated {beta}2-Adrenergic Receptor and {beta}-Arrestin. *Science*. 4:4.

Shiba, T., H. Takatsu, T. Nogi, N. Matsugaki, M. Kawasaki, N. Igarashi, M. Suzuki, R. Kato, T. Earnest, K. Nakayama, and S. Wakatsuki. 2002. Structural basis for recognition of acidic-cluster dileucine sequence by GGA1. *Nature*. 415:937-41.

Sigal, C.T., W. Zhou, C.A. Buser, S. McLaughlin, and M.D. Resh. 1994. Amino-terminal basic residues of Src mediate membrane binding through electrostatic interaction with acidic phospholipids. *Proc Natl Acad Sci U S A*. 91:12253-7.

Simmen, T., S. Honing, A. Icking, R. Tikkanen, and W. Hunziker. 2002. AP-4 binds basolateral signals and participates in basolateral sorting in epithelial MDCK cells. *Nat Cell Biol*. 4:154-9.

Simpson, F., A.A. Peden, L. Christopoulou, and M.S. Robinson. 1997. Characterization of the adaptor-related protein complex, AP-3. *J Cell Biol*. 137:835-45.

Slepnev, V.I., and P. DeCamilli. 2000. Accessory factors in clathrin-dependent synaptic vesicle endocytosis. *Nat Rev Neurosci*. 1:161-172.

Sollner, T., S.W. Whiteheart, M. Brunner, H. Erdjument-Bromage, S. Geromanos, P. Tempst, and J.E. Rothman. 1993. SNAP receptors implicated in vesicle targeting and fusion. *Nature*. 362:318-24.

Soskic, V., E. Nyakatura, M. Roos, W. Muller-Esterl, and J. Godovac-Zimmermann. 1999. Correlations in palmitoylation and multiple phosphorylation of rat bradykinin B2 receptor in Chinese hamster ovary cells. *J Biol Chem*. 274:8539-45.

Soyombo, A.A., and S.L. Hofmann. 1997a. Molecular cloning and expression of palmitoyl-protein thioesterase 2 (PPT2), a homolog of lysosomal palmitoyl- protein thioesterase with a distinct substrate specificity. *J Biol Chem*. 272:27456-27463.

Soyombo, A.A., and S.L. Hofmann. 1997b. Molecular cloning and expression of palmitoyl-protein thioesterase 2 (PPT2), a homolog of lysosomal palmitoyl-protein thioesterase with a distinct substrate specificity. *J Biol Chem*. 272:27456-63.

Stamnes, M.A., and J.E. Rothman. 1993. The binding of AP-1 clathrin adaptor particles to Golgi membranes requires ADP-ribosylation factor, a small GTP-binding protein. *Cell.* 73:999-1005.

Stauffer, T.P., S. Ahn, and T. Meyer. 1998. Receptor-induced transient reduction in plasma membrane PtdIns(4,5)P2 concentration monitored in living cells. *Curr Biol.* 8:343-6.

Steegmaier, M., J. Klumperman, D.L. Foletti, J.S. Yoo, and R.H. Scheller. 1999. Vesicle-associated membrane protein 4 is implicated in trans-Golgi network vesicle trafficking. *Mol Biol Cell.* 10:1957-1972.

Stein, M., J.E. Zijderhand-Bleekemolen, H. Geuze, A. Hasilik, and K. von Figura. 1987. Mr 46,000 mannose 6-phosphate specific receptor: its role in targeting of lysosomal enzymes. *Embo J.* 6:2677-81.

Stephens, D.J., and G. Banting. 1998. Specificity of interaction between adaptor-complex medium chains and the tyrosine-based sorting motifs of TGN38 and lgp120. *Biochem J.* 335:567-572.

Stinchcombe, J.C., D.C. Barral, E.H. Mules, S. Booth, A.N. Hume, L.M. Machesky, M.C. Seabra, and G.M. Griffiths. 2001. Rab27a is required for regulated secretion in cytotoxic T lymphocytes. *J Cell Biol.* 152:825-34.

Storch, S., and T. Braulke. 2001. Multiple C-terminal motifs of the 46-kDa mannose 6- phosphate receptor tail contribute to efficient binding of medium chains of AP-2 and AP-3. *J Biol Chem.* 276:4298-4303.

Storrie, B., and M. Desjardins. 1996. The biogenesis of lysosomes: is it a kiss and run, continuous fusion and fission process? *Bioessays.* 18:895-903.

Takatsu, H., Y. Katoh, Y. Shiba, and K. Nakayama. 2001. Golgi-localizing, gamma-adaptin ear homology domain, ADP-ribosylation factor-binding (GGA) proteins interact with acidic dileucine sequences within the cytoplasmic domains of sorting receptors through their Vps27p/Hrs/STAM (VHS) domains. *J Biol Chem.* 276:28541-5.

Takei, K., O. Mundigl, L. Daniell, and P. De Camilli. 1996. The synaptic vesicle cycle: a single vesicle budding step involving clathrin and dynamin. *J Cell Biol.* 133:1237-50.

Takeshita, T., T. Arita, H. Asao, N. Tanaka, M. Higuchi, H. Kuroda, K. Kaneko, H. Munakata, Y. Endo, T. Fujita, and K. Sugamura. 1996. Cloning of a novel signal-transducing adaptor molecule containing an SH3 domain and ITAM. *Biochem Biophys Res Commun.* 225:1035-9.

Tebar, F., S.K. Bohlander, and A. Sorkin. 1999. Clathrin assembly lymphoid myeloid leukemia (CALM) protein: Localization in endocytic-coated pits, interactions with clathrin, and the impact of overexpression on clathrin- mediated traffic. *Mol Biol Cell.* 10:2687-2702.

Tebar, F., T. Sorkina, A. Sorkin, M. Ericsson, and T. Kirchhausen. 1996. Eps15 is a component of clathrin-coated pits and vesicles and is located at the rim of coated pits. *J Biol Chem.* 271:28727-30.

Tikkanen, R., S. Obermuller, K. Denzer, R. Pungitore, H.J. Geuze, K. von Figura, and S. Höning. 2000. The dileucine motif within the tail of MPR46 is required for sorting of the receptor in endosomes. *Traffic.* 1:631-40.

Tikkanen, R., M. Peltola, C. Oinonen, J. Rouvinen, and L. Peltonen. 1997. Several cooperating binding sites mediate the interaction of a lysosomal enzyme with phosphotransferase. *Embo J.* 16:6684-6693.

Tong, P.Y., W. Gregory, and S. Kornfeld. 1989. Ligand interactions of the cation-independent mannose 6-phosphate receptor. The stoichiometry of mannose 6-phosphate binding. *J Biol Chem.* 264:7962-9.

Tong, P.Y., and S. Kornfeld. 1989. Ligand interactions of the cation-dependent mannose 6-phosphate receptor. Comparison with the cation-independent mannose 6-phosphate receptor. *J Biol Chem.* 264:7970-5.

Tong, P.Y., S.E. Tollefsen, and S. Kornfeld. 1988. The cation-independent mannose 6-phosphate receptor binds insulin-like growth factor II. *J Biol Chem.* 263:2585-8.

References

Torrisi, M.R., L.V. Lotti, F. Belleudi, R. Gradini, A.E. Salcini, S. Confalonieri, P.G. Pelicci, and P.P. Difiore. 1999. Eps15 is recruited to the plasma membrane upon epidermal growth factor receptor activation and localizes to components of the endocytic pathway during receptor internalization. *Mol Biol Cell*. 10:417-434.

Towbin, H., T. Staehelin, and J. Gordon. 1979. Electrophoretic transfer of proteins from polyacrylamide gels to nitrocellulose sheets: procedure and some applications. *Proc Natl Acad Sci U S A*. 76:4350-4.

Towler, D.A., S.P. Adams, S.R. Eubanks, D.S. Towery, E. Jackson-Machelski, L. Glaser, and J.I. Gordon. 1987. Purification and characterization of yeast myristoyl CoA:protein N-myristoyltransferase. *Proc Natl Acad Sci U S A*. 84:2708-12.

Traub, L.M., S. Kornfeld, and E. Ungewickell. 1995. Different domains of the AP-1 adaptor complex are required for Golgi membrane binding and clathrin recruitment. *J Biol Chem*. 270:4933-42.

Traub, L.M., J.A. Ostrom, and S. Kornfeld. 1993. Biochemical dissection of AP-1 recruitment onto Golgi membranes. *J Cell Biol*. 123:561-73.

Tu, Y., S. Popov, C. Slaughter, and E.M. Ross. 1999. Palmitoylation of a conserved cysteine in the regulator of G protein signaling (RGS) domain modulates the GTPase-activating activity of RGS4 and RGS10. *J Biol Chem*. 274:38260-7.

Tu, Y., J. Wang, and E.M. Ross. 1997. Inhibition of brain Gz GAP and other RGS proteins by palmitoylation of G protein alpha subunits. *Science*. 278:1132-5.

Ueno, K., and Y. Suzuki. 1997. p260/270 expressed in embryonic abdominal leg cells of Bombyx mori can transfer palmitate to peptides. *J Biol Chem*. 272:13519-26.

Umeda, A., A. Meyerholz, and E. Ungewickell. 2000. Identification of the universal cofactor (auxilin 2) in clathrin coat dissociation. *Eur J Cell Biol*. 79:336-42.

Ungewickell, E., H. Ungewickell, S.E. Holstein, R. Lindner, K. Prasad, W. Barouch, B. Martin, L.E. Greene, and E. Eisenberg. 1995. Role of auxilin in uncoating clathrin-coated vesicles. *Nature*. 378:632-5.

van't Hof, W., and M.D. Resh. 1997. Rapid plasma membrane anchoring of newly synthesized p59fyn: selective requirement for NH2-terminal myristoylation and palmitoylation at cysteine-3. *J Cell Biol*. 136:1023-35.

van't Hof, W., and M.D. Resh. 1999. Dual fatty acylation of p59(Fyn) is required for association with the T cell receptor zeta chain through phosphotyrosine-Src homology domain-2 interactions. *J Cell Biol*. 145:377-89.

Veit, M., R. Laage, L. Dietrich, L. Wang, and C. Ungermann. 2001. Vac8p release from the SNARE complex and its palmitoylation are coupled and essential for vacuole fusion. *Embo J*. 20:3145-55.

Veit, M., and M.F. Schmidt. 1993. Timing of palmitoylation of influenza virus hemagglutinin. *FEBS Lett*. 336:243-7.

Veit, M., and M.F. Schmidt. 2001. Enzymatic depalmitoylation of viral glycoproteins with acyl-protein thioesterase 1 in vitro. *Virology*. 288:89-95.

Verkruyse, L.A., and S.L. Hofmann. 1996. Lysosomal targeting of palmitoyl-protein thioesterase. *J Biol Chem*. 271:15831-6.

Vesa, J., E. Hellsten, L.A. Verkruyse, L.A. Camp, J. Rapola, P. Santavuori, S.L. Hofmann, and L. Peltonen. 1995. Mutations in the palmitoyl protein thioesterase gene causing infantile neuronal ceroid lipofuscinosis. *Nature*. 376:584-7.

Vigers, G.P., R.A. Crowther, and B.M. Pearse. 1986. Location of the 100 kd-50 kd accessory proteins in clathrin coats. *Embo J*. 5:2079-85.

Waelter, S., E. Scherzinger, R. Hasenbank, E. Nordhoff, R. Lurz, H. Goehler, C. Gauss, K. Sathasivam, G.P. Bates, H. Lehrach, and E.E. Wanker. 2001. The huntingtin interacting protein HIP1 is a clathrin and alpha-adaptin-binding protein involved in receptor-mediated endocytosis. *Hum Mol Genet*. 10:1807-17.

References

Wan, L., S.S. Molloy, L. Thomas, G. Liu, Y. Xiang, S.L. Rybak, and G. Thomas. 1998. PACS-1 defines a novel gene family of cytosolic sorting proteins required for trans-Golgi network localization. *Cell*. 94:205-16.

Ward, D.M., S.L. Shiflett, and J. Kaplan. 2002. Chediak-Higashi syndrome: a clinical and molecular view of a rare lysosomal storage disorder. *Curr Mol Med*. 2:469-77.

Waterman, H., G. Levkowitz, I. Alroy, and Y. Yarden. 1999. The RING finger of c-Cbl mediates desensitization of the epidermal growth factor receptor. *J Biol Chem*. 274:22151-4.

Waters, M.G., and S.R. Pfeffer. 1999. Membrane tethering in intracellular transport. *Curr Opin Cell Biol*. 11:453-459.

Weber, T., B.V. Zemelman, J.A. McNew, B. Westermann, M. Gmachl, F. Parlati, T.H. Sollner, and J.E. Rothman. 1998. SNAREpins: Minimal machinery for membrane fusion. *Cell*. 92:759-772.

Wendland, M., K. von Figura, and R. Pohlmann. 1991a. Mutational analysis of disulfide bridges in the Mr 46,000 mannose 6-phosphate receptor. Localization and role for ligand binding. *J Biol Chem*. 266:7132-6.

Wendland, M., A. Waheed, B. Schmidt, A. Hille, G. Nagel, K. von Figura, and R. Pohlmann. 1991b. Glycosylation of the Mr 46,000 mannose 6-phosphate receptor. Effect on ligand binding, stability, and conformation. *J Biol Chem*. 266:4598-604.

Wendler, F., L. Page, S. Urbe, and S.A. Tooze. 2001. Homotypic fusion of immature secretory granules during maturation requires syntaxin 6. *Mol Biol Cell*. 12:1699-709.

Wenk, J., A. Hille, and K. von Figura. 1991. Quantitation of Mr 46000 and Mr 300000 mannose 6-phosphate receptors in human cells and tissues. *Biochem Int*. 23:723-31.

Westcott, K.R., and L.H. Rome. 1988. Cation-independent mannose 6-phosphate receptor contains covalently bound fatty acid. *J Cell Biochem*. 38:23-33.

Wigge, P., and H.T. McMahon. 1998. The amphiphysin family of proteins and their role in endocytosis at the synapse. *Trends Neurosci*. 21:339-344.

Worby, C.A., and J.E. Dixon. 2002. Sorting out the cellular functions of sorting nexins. *Nat Rev Mol Cell Biol*. 3:919-31.

Xu, Y., H. Hortsman, L. Seet, S.H. Wong, and W. Hong. 2001. SNX3 regulates endosomal function through its PX-domain-mediated interaction with PtdIns(3)P. *Nat Cell Biol*. 3:658-66.

Yamabhai, M., N.G. Hoffman, N.L. Hardison, P.S. McPherson, L. Castagnoli, G. Cesareni, and B.K. Kay. 1998. Intersectin, a novel adaptor protein with two Eps15 homology and five Src homology 3 domains. *J Biol Chem*. 273:31401-31407.

Yamashiro, D.J., and F.R. Maxfield. 1984. Acidification of endocytic compartments and the intracellular pathways of ligands and receptors. *J Cell Biochem*. 26:231-46.

Yang, C., C.P. Spies, and R.W. Compans. 1995. The human and simian immunodeficiency virus envelope glycoprotein transmembrane subunits are palmitoylated. *Proc Natl Acad Sci U S A*. 92:9871-5.

Yeh, D.C., J.A. Duncan, S. Yamashita, and T. Michel. 1999. Depalmitoylation of endothelial nitric-oxide synthase by acyl-protein thioesterase 1 is potentiated by Ca2+- calmodulin. *J Biol Chem*. 274:33148-33154.

Yik, J.H., A. Saxena, J.A. Weigel, and P.H. Weigel. 2002. Nonpalmitoylated human asialoglycoprotein receptors recycle constitutively but are defective in coated pit-mediated endocytosis, dissociation, and delivery of ligand to lysosomes. *J Biol Chem*. 277:40844-52.

York, S.J., L.S. Arneson, W.T. Gregory, N.M. Dahms, and S. Kornfeld. 1999. The rate of internalization of the mannose 6- phosphate/insulin-like growth factor II receptor is enhanced by multivalent ligand binding. *J Biol Chem*. 274:1164-1171.

Zhang, F.L., and P.J. Casey. 1996. Protein prenylation: molecular mechanisms and functional consequences. *Annu Rev Biochem*. 65:241-69.

Zhang, W., R.P. Trible, and L.E. Samelson. 1998. LAT palmitoylation: its essential role in membrane microdomain targeting and tyrosine phosphorylation during T cell activation. *Immunity*. 9:239-46.

References

Zheng, B., Y.C. Ma, R.S. Ostrom, C. Lavoie, G.N. Gill, P.A. Insel, X.Y. Huang, and M.G. Farquhar. 2001. RGS-PX1, a GAP for GalphaS and sorting nexin in vesicular trafficking. *Science*. 294:1939-42.

Zhu, Y., B. Doray, A. Poussu, V.P. Lehto, and S. Kornfeld. 2001. Binding of GGA2 to the lysosomal enzyme sorting motif of the mannose 6-phosphate receptor. *Science*. 292:1716-8.

Zizioli, D., C. Meyer, G. Guhde, P. Saftig, K. Vonfigura, and P. Schu. 1999. Early embryonic death of mice deficient in gamma-adaptin. *J Biol Chem*. 274:5385-5390.

Die VDM Verlagsservicegesellschaft sucht für wissenschaftliche Verlage abgeschlossene und herausragende

Dissertationen, Habilitationen, Diplomarbeiten, Master Theses, Magisterarbeiten usw.

für die kostenlose Publikation als Fachbuch.

Sie verfügen über eine Arbeit, die hohen inhaltlichen und formalen Ansprüchen genügt, und haben Interesse an einer honorarvergüteten Publikation?

Dann senden Sie bitte erste Informationen über sich und Ihre Arbeit per Email an *info@vdm-vsg.de*.

Sie erhalten kurzfristig unser Feedback!

VDM Verlagsservicegesellschaft mbH
Dudweiler Landstr. 99
D - 66123 Saarbrücken
www.vdm-vsg.de

Telefon +49 681 3720 174
Fax +49 681 3720 1749

Die VDM Verlagsservicegesellschaft mbH vertritt

Printed by Books on Demand GmbH, Norderstedt / Germany